PAULI LECTURES ON PHYSICS VOLUME 5

Wave Mechanics

Wolfgang Pauli

Edited by Charles P. Enz
Translated by S. Margulies and H. R. Lewis
Foreword by Victor F. Weisskopf

DOVER PUBLICATIONS, INC.
Mineola. New York

Bibliographical Note

This Dover edition, first published in 2000, is an unabridged
republication of the work originally published in 1973 by The MIT
Press, Cambridge, Massachusetts and London, England.

Library of Congress Cataloging-in-Publication Data

Pauli, Wolfgang, 1900–1958.
    [Vorlesung über Wellenmechanik. English]
    Wave mechanics / Wolfgang Pauli ; edited by Charles P. Enz ;
translated by S. Margulies and H.R. Lewis ; foreword by Victor F.
Weisskopf.
       p. cm. — (Pauli lectures on physics ; v. 5)
    Originally published: Cambridge, Mass. : MIT Press, 1973.
    Includes bibliographical references and index.
    ISBN 0-486-41462-0 (pbk.)
    1. Wave mechanics. I. Enz, Charles P. (Charles Paul), 1925– II.
Title.

QC3 .P35 2000 vol.5
[QC174.2]
530 s—dc21
[531'.1133]

00-031580

Manufactured in the United States of America
Dover Publications, Inc., 31 East 2nd Street, Mineola, N.Y. 11501

Pauli Lectures on Physics in Dover Editions

Volume 1.   Electrodynamics (41457-4)
Volume 2.   Optics and the Theory of Electrons (41458-2)
Volume 3.   Thermodynamics and the Kinetic Theory of Gases
            (41461-2)
Volume 4.   Statistical Mechanics (41460-4)
Volume 5.   Wave Mechanics (41462-0)
Volume 6.   Selected Topics in Field Quantization (41459-0)

Contents

Foreword by Victor F. Weisskopf   ix

Preface by the Editor   xi

Preface by the Students   xiii

Introduction   xv

1. Wave Functions of Force-Free Particles   1

1. Association of waves with particles   1
2. The wave function and wave equation   3
3. The uncertainty principle   5
4. Wave packets and the mechanics of point particles.
   Probability density   10
5. Measuring arrangements. Discussion of examples   12
6. Classical statistics and quantum statistics   19

2. Description of a Particle in a Box and in Free Space   23

7. One particle in a box. The equation of continuity   23
8. Normalization in the continuum. The Dirac δ-function   27
9. The completeness relation. Expansion theorem   31
10. Initial-value problem and the fundamental solution   33

3. Particle in a Field of Force   38

11. The Hamiltonian operator   38
12. Hermitian operators   40

13. Expectation values and the classical equation of motion.
Commutation relations (commutators)   43

4. More than One Particle   51

14. More than one particle   51

5. Eigenvalue Problems. Functions of Mathematical Physics   55

15. The linear harmonic oscillator. Hermite polynomials   55
16. Matrix calculus illustrated with the linear harmonic oscillator   63
17. The harmonic oscillator in a plane. Degeneracy   72
18. The hydrogen atom   88

6. Collision Processes   107

19. Asymptotic solution of the scattering problem   108
20. The scattering cross section. The Rutherford scattering formula   110
21. Solution of the force-free wave equation   112
22. Expansion of a plane wave in Legendre polynomials   115
23. Solution of the Schrödinger equation with an arbitrary
central potential   116
24. The Born approximation   120
25. Scattering of low-energy particles   123

7. Approximate Methods for Solving the Wave Equation   126

26. Eigenvalue problem of a particle in a uniform field   126
27. The WKB method   132

8. Matrices and Operators. Perturbation Theory   138

28. General relationship between matrices and operators.
Transformation theory   138
29. General formalism of perturbation theory in the matrix
representation   143
30. Time-dependent perturbation   147

9. Angular Momentum and Spin   152

31. General commutation relations 152
32. Matrix elements of the angular momentum 154
33. Spin 156
34. Spinors and space rotations 159

10. Identical Particles with Spin 165

35. Symmetry classes 165
36. The exclusion principle 167
37. The helium atom 169
38. Collision of two identical particles: Mott's theory 173
39. The statistics of nuclear spins 175

Exercises 177

40. Fundamental solution for interval 177
41. Bound states and tunnel effect 179
42. Kronig-Penney potential 180
43. Spherical harmonics 180
44. Fundamental solution for harmonic oscillator 182
45. Angular momentum 184
46. Partial waves 185
47. The symmetrical top 186

Bibliography 191

Appendix. Comments by the Editor 193

Index 199

Foreword

It is often said that scientific texts quickly become obsolete. Why are the Pauli lectures brought to the public today, when some of them were given as long as twenty years ago? The reason is simple: Pauli's way of presenting physics is never out of date. His famous article on the foundations of quantum mechanics appeared in 1933 in the German encyclopedia *Handbuch der Physik*. Twenty-five years later it reappeared practically unchanged in a new edition, whereas most other contributions to this encyclopedia had to be completely rewritten. The reason for this remarkable fact lies in Pauli's style, which is commensurate to the greatness of its subject in its clarity and impact. Style in scientific writing is a quality that today is on the point of vanishing. The pressure of fast publication is so great that people rush into print with hurriedly written papers and books that show little concern for careful formulation of ideas. Mathematical and instrumental techniques have become complicated and difficult; today most of the effort of writing and learning is devoted to the acquisition of these techniques instead of insight into important concepts. Essential ideas of physics are often lost in the dense forest of mathematical reasoning. This situation need not be so. Pauli's lectures show how physical ideas can be presented clearly and in good mathematical form, without being hidden in formalistic expertise.

Pauli was not an accomplished lecturer in the technical sense of the word. It was often difficult to follow his courses. But when the sequence of his thoughts and the structure of his logic become apparent, the attentive follower is left with a new and deeper knowledge of essential concepts and with a clearer insight into the splendid architecture of reason, which is theoretical physics. The value of the lecture notes is not diminished by the fact that they were written not by him but by some of his collaborators. They bear the mark of the master in their conceptual structure and their mathematical rigidity. Only here and there does one miss words and comments of the master. Neither does one notice the passing of time in his lectures, with the sole exception of the lectures on field quantization, in which some concepts are formulated in a way that may appear old-fashioned to some today. But even these lectures should be of use to modern students because of their compactness and their direct approach to the central problems.

May this volume serve as an example of how the concepts of theoretical physics were conceived and taught by one of the great men who created them.

Victor F. Weisskopf

Cambridge, Massachusetts

Preface

It is astonishing that Pauli had given a complete version of
this course "Wellenmechanik" only once, during the winter
term of 1956–57. This is due to the fact that although quantum
mechanics was part of the physics diploma curriculum of ETH
at Zürich, a regular course on the subject did not exist for a
long time, a surprising fact considering that the author of the
famous article "Die allgemeinen Prinzipien der Wellenmechanik" (*Handbuch der Physik*, Band 24/1, Springer, Berlin,
1933) joined the ETH faculty in 1928.

While much of the spirit of the "Handbuchartikel" is re-
flected in this course, Pauli took the opportunity to put into it
much of his extended knowledge of that 19th-century mathematics symbolized by Whittaker and Watson's *A Course of
Modern Analysis*, for which he shared a liking with his teacher
Arnold Sommerfeld. This is one of the features of the present
volume not found in the same detail and rigor in other books
on quantum mechanics, which makes it a worthwhile text also
for today's students.

But apart from this more technical aspect it is of interest to
see how the most critical among the founders of quantum
theory actually taught his subject. As mentioned in the preface
by the students he did it mainly in mathematical language with
few but penetrating comments. And he chose the topics of this
course according to their conceptual and historical interest.
Examples are the probabilistic nature of quantum theory, the

concept of spin, the problem of identical particles, and the relation of the statistics of rotational states of diatomic molecules
to nuclear spin. For this last reason and because of Pauli's deep
involvement in the development of quantum mechanics, I have
tried in the appendix to comment on some of the fascinating
highlights of its history.

Pauli's second delivery of the course starting in October 1958
was tragically interrupted by his death in December of that
year. He had treated the first 16 sections, as well as the hydrogen atom in polar coordinates, when I was asked to take over
the course. As mentioned in the preface by the students, Pauli
had intended to correct their notes taken from the first delivery; he therefore made corrections only for the part mentioned. Some of the responsibility for the remainder of the
course and, in particular, for the wording of the problem sections therefore falls upon me.

The carefully prepared notes by Herlach and Knoepfel have
allowed a fairly smooth editing job of this English translation,
in which the work done by the translators was of great help.

Charles P. Enz

Geneva, 27 October 1971

Preface by the Students

Professor Pauli participated personally in the development of wave mechanics. For this reason he was able to deliver this lecture course of four hours per week without notes, and he did it with particular pleasure, as a complete master of the subject.

At Professor Pauli's request, we took notes in his lectures during the winter semester of 1956–1957 and worked those notes into their present form. He wanted to correct the manuscript by delivering his lectures from the notes during the winter semester of 1958–1959. Unfortunately, it was not possible for him to complete that objective. However, the very small number of corrections in the part which he looked through (up to the hydrogen atom) encourages us to publish the entire set of notes, and we hope that this will be as our revered teacher would have wished.

In these lectures he wanted particularly to treat fully the mathematical foundations of the theory. This treatment of wave mechanics differs in yet another way from that found in the usual textbooks.

Professor Pauli was fond of expressing himself with formulas and only a few words. Once he said, "One should not write so much." In working through the notes we have taken pains to preserve the originality and the special style of Professor Pauli's lectures. We have also permitted ourselves to smuggle some of his characteristic remarks into the text.

We are especially grateful to Dr. C. Enz for correcting the second part.

F. Herlach
H. E. Knoepfel

Zürich
December 1958

Introduction

The discovery of the quantum of action by Planck in 1900 initiated the development of wave mechanics. In the beginning there were many problems associated with the theory. It took all of 25 years until wave mechanics was formulated as a self-consistent theory. Of course, this could only be achieved by giving up a certain concreteness; in particular, a system of several particles can no longer be described with concrete waves. Very soon after the first papers by de Broglie, Heisenberg, and Schrödinger in 1927, the fundamental principles of the theory were logically complete. Since then, the theory has proved useful in numerous fields of physics, and it has been corroborated repeatedly by experiment.

Wave Mechanics

PAULI LECTURES ON PHYSICS VOLUME 5

Chapter 1. Wave Functions of Force-Free Particles[1]

1. ASSOCIATION OF WAVES WITH PARTICLES

A particle with energy E and momentum p can be associated with a wave, $A \exp[i(k \cdot x - \omega t)]$, as follows ($k = (2\pi/\lambda)n$ is the wave vector and n is the wave normal). For light quanta there are the relations

$$E = h\omega, \qquad p = hk, \qquad [1.1]$$

which are relativistically invariant. In addition, there are the relations

$$\left. \begin{array}{ll} |k| \equiv k = \dfrac{\omega}{c}, & |p| \equiv p = \dfrac{E}{c} \\[2mm] k^2 = \dfrac{\omega^2}{c^2}, & p^2 = \dfrac{E^2}{c^2} \end{array} \right\}. \qquad [1.2]$$

From relativistic particle mechanics we have the formulas

$$\frac{E}{c} = \sqrt{p^2 + m^2 c^2} \qquad (m = \text{rest mass}), \qquad [1.3]$$

$$E = \frac{mc^2}{\sqrt{1 - v^2/c^2}}, \qquad p = \frac{mv}{\sqrt{1 - v^2/c^2}}, \qquad [1.4]$$

where [1.3] follows from [1.4]. Because of the general for-

---

[1] Remarks:

1. In these lectures we use the symbol h to denote the quantity 1.05×10^{-34} joule·sec. In the older literature this quantity was usually denoted by $\hbar$.
2. Limits of integration $-\infty$ and $+\infty$ will usually be omitted.

mula from mechanics,

$$\mathrm{d}E = \boldsymbol{v} \cdot \mathrm{d}\boldsymbol{p}, \qquad \text{or (in components)} \qquad v_\iota = \frac{\partial E}{\partial p_\iota}, \qquad [1.5]$$

we can also derive [1.4] from [1.3]:

$$\left. \begin{aligned} v_\iota &= \frac{\partial E}{\partial p_\iota} = c\,\frac{p_\iota}{\sqrt{\boldsymbol{p}^2 + mc^2}} = c^2\,\frac{p_\iota}{E} \\ \frac{v^2}{c^2} &= \frac{c^2 p^2}{E^2}, \qquad 1 - \frac{v^2}{c^2} = 1 - \frac{c^2 p^2}{E^2} = \frac{m^2 c^4}{E^2} \\ \frac{v^2}{c^2} &= \frac{p^2}{p^2 + m^2 c^2}, \qquad p^2\left(1 - \frac{v^2}{c^2}\right) = m^2 v^2 \end{aligned} \right\}. \qquad [1.6]$$

Equations [1.5] and [1.6] can also be combined to give the relation

$$(1/c^2)E\,\mathrm{d}E = \boldsymbol{p} \cdot \mathrm{d}\boldsymbol{p}.$$

The idea of de Broglie was that [1.1] should also be valid for a material particle, in which case [1.2] must be replaced by

$$\frac{\omega}{c} = \sqrt{k^2 + \frac{m^2 c^2}{h^2}}, \qquad \frac{\omega^2}{c^2} = k^2 + \frac{m^2 c^2}{h^2}, \qquad [1.7]$$

which follows from [1.1] and [1.3]. For light ($m = 0$) we again have [1.2].

Substituting [1.1] into [1.5] yields

$$v_\iota = \frac{\partial \omega}{\partial k_\iota}; \qquad [1.8]$$

that is, *particle velocity = group velocity of the associated wave.* From [1.1] and [1.6] follows $|\boldsymbol{v}| = c^2(k/\omega)$; accordingly, for the phase velocity u,

$$u = \frac{\omega}{k} = \frac{c^2}{v}. \qquad [1.9]$$

Since $v < c$, therefore $u > c$.

2. THE WAVE FUNCTION AND WAVE EQUATION

a. *Superposition of plane waves. Wave packets*

The most general wave packet is of the form [2]

$$\psi(\boldsymbol{x}, t) = \iiint A(k_1, k_2, k_3) \exp\left[i(\boldsymbol{k}\cdot\boldsymbol{x} - \omega t)\right] dk_1 dk_2 dk_3 \qquad [2.1]$$

where ω is now given by [1.7].

This wave function ψ satisfies the relativistic scalar wave equation,

$$\left(\nabla^2 - \frac{1}{c^2}\frac{\partial^2}{\partial t^2}\right)\psi(\boldsymbol{x}, t) = \frac{m^2 c^2}{h^2}\psi(\boldsymbol{x}, t), \qquad [2.2]$$

as can be seen by substituting [2.1] into [2.2]. Because of [1.7], the wave equation is then satisfied identically. Therefore, we can also write

$$\frac{\partial}{\partial x_\iota} \sim ik_\iota, \qquad \frac{\partial}{\partial t} \sim -i\omega. \qquad [2.3]$$

These correspondences, together with [1.1], yield the important connections between the differential operators of time and space and the classical quantities $\boldsymbol{p}$ and $\boldsymbol{E}$:

$$-ih\frac{\partial}{\partial x_\iota} \sim p_\iota, \qquad ih\frac{\partial}{\partial t} \sim E. \qquad [2.4]$$

These form a translational key between the quantities of classical mechanics and the operators of wave mechanics.

b. *Transition to the nonrelativistic approximation*

In mechanics, for $v \ll c$ $(c \to \infty)$ and $p \ll mc$, we have

$$\frac{E}{c} = \sqrt{p^2 + m^2 c^2} \sim mc\left(1 + \frac{1}{2}\frac{p^2}{m^2 c^2} + \dots\right)$$
$$= \frac{1}{c}\left(mc^2 + \frac{1}{2}\frac{p^2}{m} + \dots\right). \qquad [2.5]$$

---

[2] See W. PAULI, *Lectures in Physics: Optics and the Theory of Electrons* (M.I.T. Press, Cambridge, Mass., 1972).

From [1.7] we also obtain

$$\omega = \frac{E}{h} = \frac{mc^2}{h} + \frac{h}{2m} k^2 + \dots \qquad [2.6]$$

(where $E = mc^2 + E_{\text{kin}}$, $E_{\text{kin}} = p^2/2m$). We define

$$\omega' = \frac{h}{2m} k^2, \qquad [2.7]$$

in order that

$$\omega = \frac{mc^2}{h} + \omega', \qquad [2.8]$$

and

$$\psi'(\boldsymbol{x}, t) = \iiint A(\boldsymbol{k}) \exp\left[i(\boldsymbol{k} \cdot \boldsymbol{x} - \omega' t)\right] \mathrm{d}^3 k, \qquad [2.9]$$

with which

$$\psi(\boldsymbol{x}, t) = \exp\left[-\frac{imc^2}{h} t\right] \psi'(\boldsymbol{x}, t). \qquad [2.10]$$

Substituted into [2.2], this gives

$$\nabla^2 \psi' + \frac{m^2 c^2}{h^2} \psi' + 2 \frac{im}{h} \frac{\partial \psi'}{\partial t} - \frac{1}{c^2} \frac{\partial^2 \psi'}{\partial t^2} = \frac{m^2 c^2}{h^2} \psi',$$

and the nonrelativistic wave equation,

$$\nabla^2 \psi' + i \frac{2m}{h} \frac{\partial \psi'}{\partial t} - \frac{1}{c^2} \frac{\partial^2 \psi'}{\partial t^2} = 0. \qquad [2.11]$$

follows. Aside from the imaginary coefficient, it corresponds to the equation for heat conduction. The imaginary coefficient assures that there is no special direction in time; [2.11] is invariant under the transformation $t \to -t$, $\psi' \to \psi'^*$, whereby $\psi^* \psi$ remains unchanged.[3]

From now on we shall always calculate with the primed quantities which we have introduced here; however, for the sake of simplicity, the primes will be left off. The quantities ω and ω' only differ by a constant; however, this is not an essential difference since only frequency differences are ever of importance in wave mechanics.

---

[3] We shall see later that the physically measurable quantity is not the wave fuction ψ, but only the probability density $\psi^* \psi$.

3. THE UNCERTAINTY PRINCIPLE

The kinematics of waves does not allow the simultaneous specification of the location and the exact wavelength of a wave. Indeed, one can only speak of the location of a wave in the case of a spatially localized wave packet. The number of different wavelengths contained in the Fourier spectrum increases as the wave packet becomes more localized. A relation of the form $\Delta k_i \Delta x_i >$ constant seems reasonable, and we now want to derive this relation quantitatively. For the sake of simplicity, we carry out the calculation in only one dimension; the generalization to three dimensions is immediate.

We consider a wave packet, [2.1], at a fixed time t, which we take to be $t = 0$. As is required by the Fourier integral transformation, [2.1] can be written symmetrically in x and k:

$$\psi(x) = \frac{1}{\sqrt{2\pi}} \int A(k) \exp[ikx] \, dk \,, \qquad [3.1]$$

$$A(k) = \frac{1}{\sqrt{2\pi}} \int \psi(x) \exp[-ikx] \, dx \,. \qquad [3.2]$$

Because of the symmetry of these formulas, all equations remain correct when the following substitutions are made:

$$
\begin{array}{ccccc}
\uparrow \psi & A & x & k & i \\
\downarrow A & \psi & k & x & -i.
\end{array}
\qquad [3.3]
$$

Further, the famous formula of Parseval,

$$N = \int \psi^*(x) \psi(x) \, dx = \int A^*(k) A(k) \, dk \,, \qquad [3.4]$$

is valid.

a. Average values of functions and operators. Normalization

For a normalized wave packet the normalization integral N equals unity by definition. A plane wave of infinite extent gives $N = \infty$ and, therefore, it cannot be normalized.

We define the average value of a function F to equal the following quantities:

$$\bar{F}(x) = \frac{\int F(x)\psi^*(x)\psi(x)\,\mathrm{d}x}{\int \psi^*(x)\psi(x)\,\mathrm{d}x}, \qquad \bar{F}(k) = \frac{\int F(k)A^*(k)A(k)\,\mathrm{d}k}{\int A^*(k)A(k)\,\mathrm{d}k}.$$

The quantities $\psi^*\psi$ and A^*A in these formulas have the meaning of a density. Later on, this interpretation will be better justified.

From now on we always assume that $\psi(x)$ and $A(k)$ are normalized:

$$\bar{F}(x) = \int F(x)\psi^*(x)\psi(x)\,\mathrm{d}x , \qquad [3.5]$$

$$\bar{F}(k) = \int F(k)A^*(k)A(k)\,\mathrm{d}k . \qquad [3.6]$$

By means of the translational key [3.3], we obtain the operator $\underline{F}(-i(\partial/\partial x))$ from the function $F(k)$. How to form the average value of such an operator, that is, on what functions to allow it to operate when the average value is calculated, can only be decided by more detailed considerations. The result is

$$\bar{F}(k) = \int \psi^*(x)\left[\underline{F}\left(-i\,\frac{\partial}{\partial x}\right)\cdot \psi(x)\right]\mathrm{d}x , \qquad [3.7]$$

$$\bar{F}(x) = \int A^*(k)\left[\underline{F}\left(+i\,\frac{\partial}{\partial k}\right)\cdot A(k)\right]\mathrm{d}k . \qquad [3.8]$$

Because of [3.3], [3.8] is a result of [3.7[, and vice versa. We prove [3.8] for the case $F=$ polynomial:

1. $F(x) = x$:

$$\bar{x} = \frac{1}{\sqrt{2\pi}}\int x\psi^*\,\mathrm{d}x \cdot \int A(k)\exp[ikx]\mathrm{d}k$$

$$= \frac{1}{\sqrt{2\pi}}\int \psi^*\,\mathrm{d}x \cdot \int A(k)\left[\left(-i\,\frac{\partial}{\partial k}\right)\exp[ikx]\right]\mathrm{d}k$$

$$= \frac{1}{\sqrt{2\pi}}\int i\,\frac{\partial A}{\partial k}\,\mathrm{d}k \cdot \int \psi^*\exp[ikx]\mathrm{d}x$$

$$= \int A^*(k)\left[\left(i\,\frac{\partial}{\partial k}\right)A(k)\right]\mathrm{d}k .$$

2. $F(x) = x^n$. The proof here is analogous but involves n partial integrations. Thus, formula [3.8] is proved for a polynomial. It can also be proved without difficulty for an entire function (using the Fourier integral theorem).

b. *The uncertainty relation*

We define

$$(\delta x)^2 = (x - \bar{x})^2 , \qquad (\delta k)^2 = (k - \bar{k})^2 . \qquad [3.9]$$

For the sake of simplicity, let

$$\bar{x} = 0 \qquad \text{and} \qquad \bar{k} = 0 ,$$

which indeed can be achieved by means of a simple coordinate translation. Using [3.7] and [3.8], along with partial integration, we obtain:

$$\overline{x^2} = \int A^*(k) \left[\left(-\frac{\partial^2}{\partial k^2} \right) A(k) \right] dk = + \int \frac{\partial A^*}{\partial k} \cdot \frac{\partial A}{\partial k} \, dk , \qquad [3.10]$$

$$\overline{k^2} = \int \psi^*(x) \left[\left(-\frac{\partial^2}{\partial x^2} \right) \psi(x) \right] dx = + \int \frac{\partial \psi^*}{\partial x} \cdot \frac{\partial \psi}{\partial x} \, dx . \qquad [3.11]$$

We now come to a quantitative determination of the uncertainty relation. In order to calculate $\overline{x^2} \cdot \overline{k^2}$, we start from the inequality

$$D \equiv \left| \frac{x}{2\overline{x^2}} \, \psi(x) + \frac{\partial \psi}{\partial x} \right|^2 \geqslant 0 ;$$

$$D = \frac{x^2}{4(\overline{x^2})^2} \, \psi\psi^* + \frac{x}{2\overline{x^2}} \left(\psi \frac{\partial \psi^*}{\partial x} + \psi^* \frac{\partial \psi}{\partial x} \right) + \frac{\partial \psi}{\partial x} \cdot \frac{\partial \psi^*}{\partial x}$$

$$= \frac{1}{4} \left(\frac{x}{\overline{x^2}} \right)^2 \psi\psi^* + \frac{1}{2} \cdot \frac{\partial}{\partial x} \left(\frac{x}{\overline{x^2}} \, \psi\psi^* \right) - \frac{1}{2} \cdot \frac{1}{\overline{x^2}} \cdot \psi\psi^* + \frac{\partial \psi}{\partial x} \cdot \frac{\partial \psi^*}{\partial x}$$

$$= \frac{1}{4} \cdot \frac{1}{(\overline{x^2})^2} (x^2 - 2\overline{x^2}) \psi\psi^* + \frac{1}{2} \frac{\partial}{\partial x} \left(\frac{x}{\overline{x^2}} \, \psi\psi^* \right) + \frac{\partial \psi}{\partial x} \cdot \frac{\partial \psi^*}{\partial x} ,$$

from which, by using [3.11], [3.5], and $\psi(\infty) \to 0$, it follows that

$$0 \leqslant \int D(x)\,\mathrm{d}x = -\frac{1}{4\overline{x^2}} + \overline{k^2}.$$

Thus, we have $\overline{k^2}\cdot\overline{x^2} \geqslant \frac{1}{4}$ or, more generally, $\overline{(\delta k)^2}\,\overline{(\delta x)^2} \geqslant \frac{1}{4}$. With

$$\Delta k \equiv +\sqrt{\overline{(\delta k)^2}}, \qquad \Delta x \equiv +\sqrt{\overline{(\delta x)^2}}, \qquad \Delta p \equiv \sqrt{\overline{(\delta p)^2}},$$

the uncertainty relation follows as a purely wave-kinematical law:

$$\Delta k \cdot \Delta x \geqslant \frac{1}{2}, \qquad \Delta p \cdot \Delta x \geqslant \frac{h}{2}. \qquad [3.12]$$

The equality signs in [3.12] hold only when $D = 0$:

$$\frac{\partial \psi}{\partial x} = -\frac{x}{2\overline{x^2}}\,\psi.$$

The solution of this differential equation is a Gaussian distribution which, when normalized, is

$$\psi(x) = \sqrt[4]{\frac{2a}{\pi}}\cdot \exp\left[-\,ax^2\right]; \qquad a = \frac{1}{4\overline{x^2}}. \qquad [3.13]$$

Thus, a wave packet with a Gaussian distribution is associated with the smallest uncertainty. Spectral analysis of such a packet gives a frequency spectrum which is again a Gaussian distribution:

$$A(k) = \frac{1}{\sqrt{2\pi}}\cdot\sqrt[4]{\frac{2a}{\pi}}\int \exp\left[-\,ax^2 - ikx\right]\mathrm{d}x \qquad \text{(from [3.2])}$$

$$= \frac{1}{\sqrt{2\pi}}\cdot\sqrt[4]{\frac{2a}{\pi}}\int \exp\left[-\,a\left(x + \frac{ik}{2a}\right)^2\right]\mathrm{d}x\cdot\exp\left[-\frac{k^2}{4a}\right]$$

$$= \frac{1}{\sqrt{2\pi}}\cdot\sqrt[4]{\frac{2a}{\pi}}\cdot\sqrt{\frac{\pi}{a}}\cdot\exp\left[-\frac{k^2}{4a}\right],$$

$$A(k) = \frac{1}{\sqrt[4]{2\pi a}}\cdot\exp\left[-\frac{k^2}{4a}\right]. \qquad [3.14]$$

c. Behavior of a wave packet in time ($t \neq 0$)

We start with the time-dependent form of the Fourier integral representation (in [3.1], $kx \rightarrow (kx - \omega t)$, ω given by [2.7]):

$$\psi(x, t) = \frac{1}{\sqrt{2\pi}} \int A(k) \exp\left[i\left(kx - \frac{hk^2}{2m}t\right)\right]dk$$

$$= \frac{1}{\sqrt{2\pi}} \int A(k, t) \exp[ikx]dk, \qquad [3.15]$$

$$A(k, t) = A(k) \cdot \exp\left[-i\frac{hk^2}{2m}t\right]$$

$$= \frac{1}{\sqrt{2\pi}} \int \psi(x, t) \exp[-ikx]dx \text{ ,} \qquad [3.16]$$

$$|A(k, t)|^2 = |A(k)|^2. \qquad [3.17]$$

Now we choose $A(k)$ to be the Gaussian distribution [3.14]:

$$\psi(x, t) = \frac{1}{\sqrt{2\pi}} \cdot \frac{1}{\sqrt[4]{2\pi a}} \cdot \int \exp\left[-\frac{k^2}{4a}\left(1 + \frac{2ihat}{m}\right) + ikx\right]dk.$$

With

$$\alpha(t) = \frac{a}{1 + 2ihat/m}$$

and

$$\int \exp\left[-\frac{k^2}{4\alpha} + ikx\right]dk = \int \exp\left[-\frac{1}{4\alpha}(k - 2ix\alpha)^2\right]dk$$

$$\times \exp[-\alpha x^2] = 2\sqrt{\alpha\pi} \exp[-\alpha x^2],$$

it follows that

$$\psi(x, t) = \sqrt[4]{\frac{2a}{\pi}} \cdot \frac{1}{\sqrt{1 + 2ihat/m}} \cdot \exp[-\alpha(t)x^2]. \qquad [3.18]$$

(For $t = 0$, we again obtain [3.13].) Hence

$$|\psi(x, t)|^2 = \sqrt{\frac{2a}{\pi}} \cdot \frac{1}{\sqrt{1 + (2aht/m)^2}} \cdot \exp[-(\alpha + \alpha^*)x^2];$$

$$\alpha + \alpha^* = \frac{2a}{1 + (2aht/m)^2} = \beta. \qquad [3.19]$$

From probability theory we know the variance of a Gaussian distribution:

$$W(x) = |\psi(x)|^2 = \sqrt{\frac{\beta}{\pi}} \exp[-\beta x^2] \quad \text{implies} \quad \overline{x^2} = \frac{1}{2\beta}.$$

Using this, we obtain

$$(\Delta x)^2 = \overline{x^2} = \frac{1}{4a}\left[1 + \left(\frac{2aht}{m}\right)^2\right].$$

Similarly, it follows that $(\Delta k)^2 = a$, with $|A(k)|^2$ given by [3.14]. Therefore,

$$(\Delta x)^2 = \frac{1}{4(\Delta k)^2} + \frac{h^2(\Delta k)^2}{m^2}\, t^2. \qquad [3.20]$$

The variance of a moving wave packet increases quadratically with time. Not only does this hold for a Gaussian distribution, it is also generally true. With $A(k, t)$, we now have

$$\overline{x^2} = +\int \frac{\partial A^*(k, t)}{\partial k} \cdot \frac{\partial A(k, t)}{\partial k}\, dk = \int \left|\frac{\partial A(k, t)}{\partial k}\right|^2 dk, \qquad [3.21]$$

in analogy with [3.10]. From [3.16] we see that

$$\frac{\partial A(k, t)}{\partial k} = \exp\left[-i\frac{hk^2}{2m}t\right] \cdot \left(\frac{\partial A(k)}{\partial k} - i\frac{hk}{m}t \cdot A(k)\right). \qquad [3.22]$$

Therefore,

$$\overline{x^2}(t) = \int \left|\frac{\partial A}{\partial k} - i\frac{hk}{m}t \cdot A\right|^2 dk = \int \frac{\partial A^*}{\partial k} \cdot \frac{\partial A}{\partial k}\, dk$$
$$+ \frac{iht}{m}\int k\left(A^*\frac{\partial A}{\partial k} - A\frac{\partial A^*}{\partial k}\right) dk + \frac{h^2}{m^2}\overline{k^2}t^2. \qquad [3.23]$$

4. WAVE PACKETS AND THE MECHANICS OF POINT PARTICLES. PROBABILITY DENSITY

We introduce the momentum p in place of the wave number k: $p = hk$.

$$\left.\begin{aligned}\varphi(p, t) &= \frac{1}{\sqrt{h}}\, A(k, t)\\[2mm]\varphi(p) &= \frac{1}{\sqrt{h}}\, A(k)\end{aligned}\right\}, \qquad [4.1]$$

$$|\varphi(p, t)|^2\, dp = |A(k, t)|^2\, dk. \qquad [4.2]$$

Equations [3.2], [3.1], and [3.4] are replaced by

$$\varphi(p) = \frac{1}{\sqrt{2\pi\hbar}} \int \psi(x) \exp\left[-\frac{i}{\hbar} px\right] dx, \qquad [4.3]$$

$$\psi(x) = \frac{1}{\sqrt{2\pi\hbar}} \int \varphi(p) \exp\left[\frac{i}{\hbar} px\right] dp, \qquad [4.4]$$

and

$$\int |\varphi(p)|^2 \, dp = \int |\psi(x)|^2 \, dx = 1, \qquad [4.5]$$

respectively.

We first remark that the relation between wave packets and the mechanics of point particles can be only of a statistical nature. A measuring arrangement defines a state. For a state at time t, the wave functions $\psi(x)$ and $\varphi(p)$ are given. However, these are not physically measurable quantities; we can only measure the probability $W(x)dx$ of finding a particle between x and $x+dx$. We call $W(x)$ the *probability density*; in terms of it we can formulate the basic assumption of wave mechanics:

$$W(x) = |\psi(x)|^2,$$

the probability for the interval between x and $x+dx$ is

$$W(x)\,dx; \qquad [4.6]$$

$$W(p) = |\varphi(p)|^2,$$

the probability for the interval between p and $p+dp$ is

$$W(p)\,dp. \qquad [4.7]$$

With these functions we can write the expectation values as (compare with [3.5] and [3.6]):

$$\overline{f(x)} = \langle f(x) \rangle_a = \int f(x)\,W(x)\,dx, \qquad [4.8]$$

$$\overline{g(p)} = \langle g(p) \rangle_a = \int g(p)\,W(p)\,dp. \qquad [4.9]$$

The reason for introducing nonmeasurable quantities ψ and φ is found in the fact that these quantities obey simple mathematical laws, particularly the linear superposition principle: If $\psi_1(x)$ and $\psi_2(x)$ denote possible states, then the state described by $C_1\psi_1(x) + C_2\psi_2(x)$ is also possible. On the other hand, the probabilities—just as optical intensities—are not additive (cross terms). Rather, they show the well-known interference effects on the observation of which wave mechanics is actually based.

Of especial note is the complete symmetry of the theory in $W(x)$ and $W(p)$; that is, all formulas remain correct when the following substitutions are made:

$$\begin{array}{ccc} \uparrow p & x & i \\ \downarrow x & p & -i. \end{array} \qquad [4.10]$$

The completeness of the symmetry comes from the fact that the probability, as a real quantity, does not change as a result of complex conjugation.

"Any speculation which destroys this symmetry by introducing intuitive pictures is not to be taken seriously." This opinion was also shared by Einstein. He believed this statistical description to be indeed correct but not complete. However, up to the present day, possibilities for extending the theory have not been found, even though no proof of the impossibility of extending the theory has been given [A-1].[4]

5. MEASURING ARRANGEMENTS. DISCUSSION OF EXAMPLES

If we consider a current of particles which passes through two small holes in a diaphragm, then we discover that the probability of finding a particle behind the diaphragm is a typical diffraction pattern, as in optics (Fig. 5.1). This

---

[4] Comments [A-1]–[A-5] appear in the Appendix on pp. 193–197.

interference of the probability is independent of intensity, that is, independent of the density of the particle current. It depends only on the locations of the holes; the holes define for us a "state" with its associated wave function.

Figure 5.1

By measuring the position and momentum of particles, we can determine only statistical properties of this state which was fixed by the experimental arrangement. This statistical description prevents contradictions between wave mechanics and the mechanics of point particles; however, it leads to an uncertainty that is characteristic of wave mechanics. If we consider, for example, the "state" represented by a single, force-free atom, we find that this state is completely changed by every measurement of position or momentum. If we measure the position of the particle, for example, then an indeterminable momentum will be transferred to the particle during the process; this will make a precise prediction of the position of the particle at a later time impossible. Although the trajectory of a celestial body can be determined more and more accurately by making successive measurements to which the laws of classical mechanics are applied (Fig. 5.2), every measure-

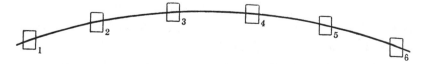

Figure 5.2

ment on an elementary particle throws the particle out of its trajectory (Fig. 5.3); that is, earlier measurements of position are useless for a further determination of the trajectory.

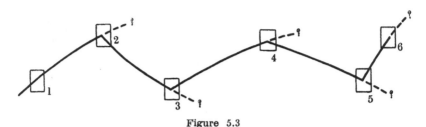

Figure 5.3

Try as one will to think of experimental arrangements and measurements, there is always an uncertainty present which is given by the Heisenberg uncertainty relation [3.12]:

$$(\Delta p)^2 \cdot (\Delta x)^2 \geqslant \frac{h^2}{4}.$$

(In Section 3 this was derived purely from wave kinematics.) Following Bohr, we call two quantities complementary if, like p and x, they satisfy an uncertainty relation. For example, E and t also are complementary.

We now want to show by means of three examples how the uncertainty relation actually works. At the beginning we can remark that the conservation laws for energy and momentum are presently considered to be firmly established experimentally and theoretically even for a single process. Likewise, it is firmly established that intensity has no influence on the interference phenomena.

a. First example: Position measurement in a microscope

From the theory of the microscope we know that convergent light is necessary for making a position measurement in a microscope (Fig. 5.4). The limit of accuracy of such a position measurement is given by Abbé's sine condition,

$$\Delta x \sim \frac{\lambda}{\sin \varepsilon}.$$

In this way we could in principle measure x arbitrarily accurately by simply making λ small. However, we must require that the microscope function classically. That is, we must require that we have a large number of light

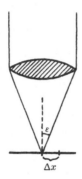

Figure 5.4

quanta, so that at least one quantum is certainly scattered from the object; we can then observe this quantum with macroscopic means (eye, photographic plate, etc.). As a result of the scattering, an indeterminate momentum is transferred to the object because, unfortunately, without disrupting the microscope, we cannot determine which path in the microscope the light quantum took. The region available to the light quantum is included in the angle ε. From this, using

$$|p| = \frac{h\nu}{c} = \frac{h}{\lambda},$$

it follows that the uncertainty in momentum is

$$\Delta p_x \sim \frac{h}{\lambda} \sin \varepsilon$$

and it is really true that

$$\Delta p_x \cdot \Delta x \sim h.$$

b. Second example: Momentum measurement using the Doppler effect

We allow a wave train of light of finite length L to impinge on the particle whose velocity v_x is to be measured (Fig. 5.5). Again, the wave train will contain so many

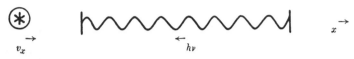

Figure 5.5

light quanta that certainly one of them will be scattered from the particle. We suppose that this light quantum approaches along the negative x direction and is scattered along the positive x direction, and we set up the energy and momentum balance for this situation before and after the collision:

momentum

$$-\frac{h\nu}{c}+p_x=\frac{h\nu'}{c}+p_x', \qquad p_x'=p_x-\frac{h\nu}{c}-\frac{h\nu'}{c} \qquad [5.1]$$

energy

$$h\nu + E = h\nu'+ E', \qquad h\nu'= h\nu + E - E'. \qquad [5.2]$$

Using the general relations (valid even relativistically)

$$\frac{\partial E}{\partial p_x} = v_x, \qquad \frac{\partial E'}{\partial p_x'} = v_x', \qquad [5.3]$$

and differentiating [5.2] with respect to p_x, we find

$$h\frac{\partial \nu'}{\partial p_x} = v_x - v_x'\cdot\frac{\partial p_x'}{\partial p_x}. \qquad [5.4]$$

Since ν is fixed, differentiation of [5.1] with respect to p_x results in

$$\frac{\partial p_x'}{\partial p_x} = 1 - \frac{h}{c}\frac{\partial \nu'}{\partial p_x}; \qquad [5.5]$$

substituting this into [5.4] gives

$$h \frac{\partial v'}{\partial p_x} = v_x - v'_x \left(1 - \frac{h}{c} \frac{\partial v'}{\partial p_x} \right)$$

or

$$h \frac{\partial v'}{\partial p_x} \left(1 - \frac{v'_x}{c} \right) = v_x - v'_x. \qquad [5.6]$$

Written in terms of uncertainties with

$$\Delta v' = (\partial v' / \partial p_x) \cdot \Delta p_x$$

we find

$$h \Delta v' = \frac{v_x - v'_x}{1 - v'_x / c} \Delta p_x \sim (v_x - v'_x) \Delta p_x. \qquad [5.7]$$

The wave train passes the particle in a finite time T; an uncertainty in v',[5]

$$\Delta v' \sim \frac{1}{T},$$

arises from this temporal limitation on the interaction. Thus,

$$\Delta p_x \sim \frac{h}{(v_x - v'_x)T}. \qquad [5.8]$$

An uncertainty in position comes about because, without ruining the arrangement for measuring momentum, we cannot determine when the particle changed its velocity during the time interval T:

$$\Delta x \sim (v_x - v'_x)T. \qquad [5.9]$$

Taken together with [5.8], this gives

$$\Delta x \cdot \Delta p_x \sim h.$$

which we had already obtained.

---

[5] This formula can be obtained, for example, by Fourier analysis of the wave packet under consideration. Moreover, it is identical to the uncertainty relation for the energy, $\Delta E \cdot \Delta t \sim h$.

c. Third example: Coherence property of light

We consider a diffraction experiment of light falling on a diaphragm with two holes (Fig. 5.6). For interference

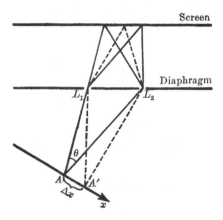

Figure 5.6

to occur, classical wave theory requires coherence of the light. What is the situation when we consider the process quantum mechanically and take a single luminous atom as the light source?

The interference pattern on the screen is a system of bright and dark bands which depend on the path difference

$$D = \frac{2\pi}{\lambda} (\overline{AL_2} - \overline{AL_1}) . \qquad [5.10]$$

If we want to determine through which hole in the diaphragm an emitted light quantum went, then we can do it by, for example, measuring the recoil momentum of the atom. To that end, we must know the momentum of the atom in the x direction to a precision

$$\Delta p_x < 2 \frac{h}{\lambda} \sin \frac{\vartheta}{2} . \qquad [5.11]$$

As a result of the wave nature of the atom, this implies

an uncertainty in position

$$\Delta x > \frac{h}{2} \cdot \frac{1}{\Delta p_x} > \frac{h}{2} \cdot \frac{\lambda}{2h \sin \vartheta/2} = \frac{\lambda}{4 \sin \vartheta/2} \cdot \qquad [5.12]$$

As far as we know, the atom could just as well be at position A'. There the path difference is

$$D' = \frac{2\pi}{\lambda} (\overline{A'L_2} - \overline{A'L_1}) , \qquad [5.13]$$

and we have

$$D - D' = \frac{2\pi}{\lambda} \cdot 2 \sin \frac{\vartheta}{2} \cdot \Delta x > \pi . \qquad [5.14]$$

This means that the interference pattern is washed out as soon as we have determined through which hole the light quantum passed. As always, the uncertainty relation prevents a contradiction between the wave and corpuscle descriptions.

6. CLASSICAL STATISTICS AND QUANTUM STATISTICS

Classically, we can say that there exists a probability density $W(p, x)$ from which, by integrating, we obtain

$$W(p) = \int W(p, x) \mathrm{d}x , \qquad \widetilde{W}(x) = \int W(p, x) \mathrm{d}p , \qquad [6.1]$$

where, however, $W(p)$ and $\widetilde{W}(x)$ are different functions. The normalization is the usual one:

$$\iint W(p, x) \, \mathrm{d}p \, \mathrm{d}x = 1, \quad \text{or} \quad \int \widetilde{W}(x) \, \mathrm{d}x = 1 , \quad \int W(p) \mathrm{d}p = 1 .$$

In this context, a measurement means a reduction of the probability, that is, a decomposition of $W(p, x)$ into parts which represent subsystems for which the probabilities are additive:

$$W(p, x) = g_1 W_1(p, x) + g_2 W_2(p, x) \qquad (0 < g < 1),$$

and also, of course,

$$W(p) = g_1 W_1(p) + g_2 W_2(p),$$
$$\widetilde{W}(x) = g_1 \widetilde{W}_1(x) + g_2 \widetilde{W}_2(x).$$

Classically, we can continue this decomposition until x and p are known to lie in the intervals $(x, x+\Delta x)$ and $(p, p+\Delta p)$, respectively, in which case we obtain the simplest distribution:

$$\widetilde{W}(x) = 0 \qquad \text{outside of} \qquad (x, \Delta x + x)$$
$$W(p) = 0 \qquad \text{outside of} \qquad (p, p+\Delta p).$$

Here there is absolutely no connection between Δp and Δx (no uncertainty relation: $h = 0$). However, we can derive a formula for the change of the probability in the course of time, for example, for a force-free mass point with statistically distributed initial conditions:

$$x = x_0 + vt = x_0 + \frac{p}{m} t \qquad \left(v = \frac{p}{m} \right).$$

With the simplifying assumptions

$$\overline{\delta x_0} = 0, \qquad \overline{\delta x_0 \, \delta p} = 0,$$

we obtain from

$$\delta x = \delta x_0 + \frac{t}{m} \delta p$$

that

$$(\Delta x)^2 = (\Delta x_0)^2 + \frac{t^2}{m^2} (\Delta p)^2.$$

This is exactly formula [3.20] which was derived on the basis of quantum theory. The only difference is that there is no connection between Δp and Δx in classical mechanics.

The specific differences between classical mechanics and quantum mechanics are

1. the uncertainty relation, and
2. the interference of probabilities.

It is these two points which lead to complications when a reduction of the probability is to be made. A state characterized by a ψ function cannot be decomposed into subsystems for which the probability is additive. Whereas in classical mechanics a measurement signifies decomposition of the probability into parts which represent subsystems, in quantum mechanics each measurement gives a new state instead of a selection of substates.

In quantum mechanics we can distinguish between two cases:

Pure case (Fig. 6.1): The probabilities are quadratic forms of ψ functions. This is true, for example, in the case of the diaphragm with two holes, both of which are definitely open. Then it is indeterminate through which of the holes in the diaphragm a particle went.

Now we introduce a shutter which always covers one of the holes, but we assume that we do not know which of the holes is being covered at any given instant. In this case, we say that it is not known through which hole the particle went. We call this case a mixture.

Mixture (Fig. 6.2): The probabilities are additive; the state cannot be described by a ψ function. Therefore, a decomposition into subsystems is possible until the state repre-

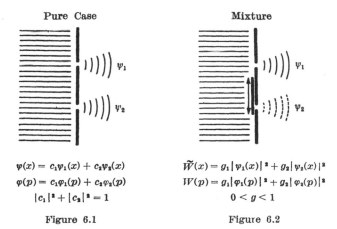

Pure Case

$$\psi(x) = c_1\psi_1(x) + c_2\psi_2(x)$$
$$\varphi(p) = c_1\varphi_1(p) + c_2\varphi_2(p)$$
$$|c_1|^2 + |c_2|^2 = 1$$

Figure 6.1

Mixture

$$\widetilde{W}(x) = g_1|\psi_1(x)|^2 + g_2|\psi_2(x)|^2$$
$$W(p) = g_1|\varphi_1(p)|^2 + g_2|\varphi_2(p)|^2$$
$$0 < g < 1$$

Figure 6.2

sented by each subsystem is a pure case; that is, in the case of a mixture, the probability is a sum of probabilities for pure cases. A mixture is always obtained if one averages over the phases, or if the relationship among the phases is destroyed in some other way.

In principle, of course, it is also possible to distinguish between a pure case and a mixture in classical mechanics, except that the pure case is then trivial in the sense that p and x are known exactly.

Chapter 2. Description of a Particle in a Box and in Free Space

7. ONE PARTICLE IN A BOX. THE EQUATION OF CONTINUITY

We first restrict our attention to a one dimensional case (Fig. 7.1).

Figure 7.1

We find the corresponding wave function by solving the wave equation

$$\frac{\partial \psi}{\partial t} = \frac{ih}{2m} \cdot \frac{\partial^2 \psi}{\partial x^2}, \qquad [7.1]$$

with boundary conditions

$$\psi(0, t) = \psi(L, t) = 0 . \qquad [7.2]$$

First, using standing waves as a starting point,

$$\psi(x, t) = u(x) \exp\left[-\frac{i}{h} Et\right], \qquad E = \frac{p^2}{2m}, \qquad [7.3]$$

we determine the solution for a stationary state. Then the probability

$$W(x) = |\psi(x, t)|^2 = |u(x)|^2$$

is independent of time. The most general solution of

$$E \cdot u = -\frac{h^2}{2m} \cdot \frac{d^2 u}{dx^2}, \qquad \text{or} \qquad u(x) + \frac{h^2}{p^2} u''(x) = 0,$$

is

$$u(x) = A \exp\left[i\frac{px}{h}\right] + B \exp\left[-i\frac{px}{h}\right].$$

Satisfaction of the boundary conditions:

$$\psi(0, t) = 0: \; B = -A, \; \text{ or } \; u = C \sin\frac{px}{h};$$

$$\psi(L, t) = 0: \; \frac{|p|L}{h} = n\pi, \; \text{ or } \; u_n = C \sin\left(\pi\frac{x}{L}n\right),$$

$$n = 0, 1, 2, \dots.$$

Normalization:

$$\int_0^L |u_n|^2 dx = |C|^2 \cdot \tfrac{1}{2}L = 1, \qquad |C| = \sqrt{\frac{2}{L}},$$

$$\psi_n(x, t) = \sqrt{\frac{2}{L}} \sin\left(\pi\frac{x}{L}n\right) \exp\left[-\frac{i}{h}E_n t\right], \; E_n = \frac{n^2\pi^2 h^2}{2mL^2}. \quad [7.4]$$

Probability density in configuration space:

$$W(x) = \frac{2}{L} \sin^2\left(\pi\frac{x}{L}n\right). \qquad [7.5]$$

Probability density in momentum space: The quantity $|p_n| = n\pi h/L$ is fixed; therefore,

$$W(p_n) = W(-p_n) = \tfrac{1}{2}. \qquad [7.6]$$

Because of this one cannot speak of the motion of a particle in a stationary state. Motion is only possible if there is a linear superposition of stationary states (wave packet):

$$\psi(x, t) = \sqrt{\frac{2}{L}} \sum_n c_n \sin\left(\pi\frac{x}{L}n\right) \exp\left[-\frac{ih\pi^2}{2mL^2}n^2 t\right]. \qquad [7.7]$$

As can be easily verified, the orthogonality relation

$$\int_0^L u_n^*(x) u_m(x)\, \mathrm{d}x = \delta_{nm} \qquad [7.8]$$

holds.

In a three-dimensional region we have an analogous eigenvalue problem (Fig. 7.2):

$$\left.\begin{aligned}
\frac{\partial \psi}{\partial t} &= \frac{ih}{2m}\nabla^2\psi, \qquad \psi = 0 \quad \text{on } S \\
\psi &= \sum_n c_n \exp\left[-\frac{i}{h}E_n t\right] u_n(x) \\
\nabla^2 u_n &+ \lambda_n u_n = 0, \qquad \lambda_n = \frac{2mE_n}{h^2}
\end{aligned}\right\} . \qquad [7.9]$$

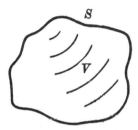

Figure 7.2

At this point we want to show that the equation of continuity, in the form known to us from electrodynamics, follows from the wave equation.[1] Here the probability current density appears in place of the electric current density,[2]

$$i = \frac{h}{2mi}\left(\psi^* \operatorname{grad}\psi - \psi \operatorname{grad}\psi^*\right), \qquad [7.10]$$

[1] Of course, here the equation of continuity is more generally valid. It includes the equation of continuity of electrodynamics as a special case.

[2] Since we can only specify the probability of finding a particle in a given volume element, a "current" of this probability appears in place of a particle current.

and the equation of continuity is

$$\frac{\partial}{\partial t}\int_V \psi^*\psi\, dV + \int_S \boldsymbol{i}\cdot d\boldsymbol{S} = 0 \left.\vphantom{\begin{array}{c}a\\a\end{array}}\right\}$$

or

$$\frac{\partial}{\partial t}(\psi^*\psi) + \operatorname{div}\boldsymbol{i} = 0 \left.\vphantom{\begin{array}{c}a\\a\end{array}}\right\} \qquad [7.11]$$

If we multiply the equation [7.9] for ψ by ψ^* and the equation for ψ^* by ψ, the sum of the resulting equations is

$$\frac{\partial}{\partial t}(\psi^*\psi) + \frac{h}{2mi}(\psi^*\,\nabla^2\psi - \psi\,\nabla^2\psi^*) = 0\;.$$

The equation of continuity follows from this equation by way of the formula due to Green and Gauss,

$$\operatorname{div}(a\operatorname{grad}b) = \operatorname{grad}a\cdot\operatorname{grad}b + a\,\nabla^2 b\;.$$

In our boundary value problem, $\boldsymbol{i} = 0$ on S; therefore, it follows that

$$\int_V \psi^*\psi\, dV = \text{constant}\;.$$

For the eigenfunctions we have the orthogonality relation

$$\int_V u_m^* u_m\, dV = 0\;, \qquad \lambda_n \neq \lambda_m\;. \qquad [7.12]$$

Proof: If we multiply the equation [7.9] for u_n by u_m^*, and the equation for u_m^* by $-u_n$, the sum of the resulting equations is

$$(u_m^*\,\nabla^2 u_n - u_n\,\nabla^2 u_m^*) + (\lambda_n - \lambda_m^*)u_m^* u_n = 0\;.$$

Using Green's formula again and applying the boundary

conditions, we obtain

$$(\lambda_n - \lambda_m^*) \int_V u_n^* u_m \, dV = 0 \; ;$$

$n = m$: $\qquad\qquad \lambda_n = \lambda_n^*$, or λ_n real,

$\lambda_n \neq \lambda_m$: $\qquad\qquad \int_V u_n^* u_m \, dV = 0$,

$\lambda_n = \lambda_m$, $n \neq m$: This is called a degeneracy of the states n and m.

The degree of degeneracy is defined as the number of independent solutions belonging to the eigenvalue λ_n. These solutions are not necessarily orthogonal. However, by an appropriate choice of basis for the eigenfunctions, they can be made orthogonal.

8. NORMALIZATION IN THE CONTINUUM. THE DIRAC δ-FUNCTION

If there are no walls or force fields present, then there is no longer a restriction on the wave function; that is, p varies continuously. We can represent $\psi(x, t)$ by a Fourier integral (see [4.1] and [3.15], with $p = hk$):

$$\psi(x, t) = \frac{1}{\sqrt{2\pi h}} \int \varphi(p, t) \exp\left[\frac{i}{h} px\right] dp$$

$$= \frac{1}{\sqrt{2\pi h}} \int \varphi(p) \exp\left[-\frac{i}{h} \frac{p^2}{2m} t + \frac{i}{h} px\right] dp \; . \qquad [8.1]$$

If we compare this to the linear superposition of eigenstates (see [7.7]),

$$\psi(x, t) = \sum_n c_n u_n(x) \exp\left[-\frac{i}{h} \frac{p_n^2}{2m} t\right],$$

then it is a simple generalization to introduce a function

$u(p, x)$ in analogy to the $u_n(x)$, in which case we have

$$\left.\begin{aligned}
\psi(x, t) &= \int \varphi(p) u(p, x) \exp\left[-\frac{i}{\hbar}\frac{p^2}{2m}t\right] \mathrm{d}p \\
&= \int \varphi(p, t) u(p, x)\, \mathrm{d}p \\
u(p, x) &= \frac{1}{\sqrt{2\pi\hbar}} \exp\left[\frac{i}{\hbar}px\right]
\end{aligned}\right\} . \qquad [8.2]$$

We write the orthogonality relation formally as

$$\int u^*(p, x) \cdot u(p', x)\, \mathrm{d}x = \delta(p - p') . \qquad [8.3]$$

This is a convenient symbolic notation; the symbol $\delta(p-p')$ (Dirac δ-function) has the following meaning:

$$\int_{p_1}^{p_2} f(p)\delta(p - p')\, \mathrm{d}p$$

$$= \begin{cases} f(p'), & \text{if } p' \text{ is in the interval } (p_1, p_2) \\ 0, & \text{if } p' \text{ is not in the interval } (p_1, p_2) . \end{cases} \qquad [8.4]$$

The "function" $\delta(x)$ does not exist in the proper sense [A-2], because it would have to have the properties

$$\delta(x) = \begin{cases} 0, & \text{for } x \neq 0 , \\ \infty, & \text{for } x = 0 . \end{cases}$$

Rather, it is the symbol for a limiting process which is carried out in the following way. We consider a function $\delta_n(x)$ such that

$$\int \delta_n(x)\, \mathrm{d}x = 1 .$$

For example:

1.
$$\delta_n(x) = n \exp[-\pi n^2 x^2]$$

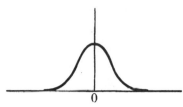

Figure 8.1

2.
$$\delta_n(x) = \frac{\sin nx}{\pi x}$$

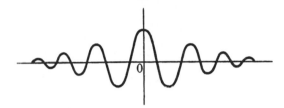

Figure 8.2

3.
$$\delta_n(x) = \begin{cases} 0, & \text{for } |x| > 1/2n, \\ n, & \text{for } |x| < 1/2n. \end{cases}$$

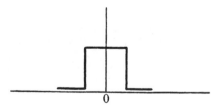

Figure 8.3

Then we have

$$\lim_{n \to \infty} \int_{p_1}^{p_2} f(p)\delta_n(p - p')\,\mathrm{d}p$$
$$= \begin{cases} f(p'), & \text{if } p' \text{ is in the interval } (p_1, p_2), \\ 0, & \text{if } p' \text{ is not in the interval } (p_1, p_2). \end{cases}$$

We can also say that $\delta(x)$ can be approximated by means of proper functions $\delta_n(x)$; for example, we can allow the rectangle to become taller and narrower indefinitely, in such a way that its area remains constant. More generally,

$$\delta_n(x) = \frac{1}{2\pi} \int A_n(p) \exp[ipx]\,dp \quad \text{with} \quad \int \delta_n(x)\,dx = 1;$$

then

$$A_n(p) = \int \delta_n(x) \exp[-ipx]\,dx, \quad A_n(0) = 1, \quad \lim_{n \to \infty} A_n(p) = 1,$$

thus

$$\delta(x) \sim \frac{1}{2\pi} \int\limits_{-\infty}^{+\infty} \exp[ipx]\,dp.$$

This means that one symbolically writes

$$f(0) = \int f(x)\delta(x)\,dx$$

in place of the exact Fourier formula

$$f(0) = \frac{1}{2\pi} \int dp \int f(x) \exp[ipx]\,dx.$$

The generalization of the δ-function to three dimensions is simple:

$$\delta^{(3)}(\boldsymbol{x} - \boldsymbol{x}') = \delta(x_1 - x_1') \cdot \delta(x_2 - x_2') \cdot \delta(x_3 - x_3').$$

Integration of [8.3] over p yields

$$\int dx \int\limits_{p_1}^{p_2} u^*(p, x)\,dp \cdot u(p', x)$$

$$= \begin{cases} 1, & \text{for} \quad p' \text{ in the interval } (p_1, p_2), \\ 0, & \text{for} \quad p' \text{ not in the interval } (p_1, p_2). \end{cases} \quad [8.5]$$

This expression has the same meaning as [8.3] and [8.4].

Now, we can also demonstrate the correctness of [8.3] because the $u(p, x)$ given in [8.2] satisfies [8.5] and, therefore, the orthogonality relation identically. That is, with

$$\frac{1}{\sqrt{2\pi h}} \int_{p_1}^{p_2} \exp\left[-\frac{i}{h} px\right] dp$$

$$= -\frac{1}{\sqrt{2\pi h}} \frac{h}{ix} \left(\exp\left[-\frac{i}{h} p_1 x\right] - \exp\left[-\frac{i}{h} p_2 x\right]\right),$$

it follows from [8.2] and [8.5] that

$$\frac{1}{2\pi} \int dx \, \frac{\exp\left[(i/h)(p'-p_1)x\right] - \exp\left[(i/h)(p'-p_2)x\right]}{ix}$$

$$= \frac{1}{2\pi} \int \frac{\sin\{(1/h)(p'-p_1)x\} - \sin\{(1/h)(p'-p_2)x\}}{x} dx$$

$$= \begin{cases} 1, & \text{for } p' \text{ in the interval } (p_1, p_2), \\ 0, & \text{for } p' \text{ not in the interval } (p_1, p_2). \end{cases}$$

This is correct because

$$\frac{1}{\pi} \int \frac{\sin ax}{x} dx = \begin{cases} +1, & \text{for } a > 0, \\ -1, & \text{for } a < 0, \end{cases}$$

$$\frac{1}{\pi} \int \frac{\cos ax - \cos bx}{x} dx = 0.$$

9. THE COMPLETENESS RELATION. EXPANSION THEOREM

Let $f(x)$ be a function such that $\int |f(x)|^2 dx$ exists. If we have a complete orthonormal set of functions $\{u_n(x)\}$, that is,

$$\int u_n^*(x) u_m(x) \, dx = \delta_{nm}, \qquad [9.1]$$

then we can make a series expansion

$$f(x) = \sum_n a_n u_n(x), \qquad [9.2]$$

in which

$$a_n = \int f(x) u_n^*(x) \, dx. \qquad [9.3]$$

We require convergence in the mean for the series in [9.2]. This is a weakened convergence requirement because possible "spikes" are integrated over:

$$\lim_{N \to \infty} \int |f(x) - \sum_{n=1}^{N} a_n u_n(x)|^2 \, dx = 0 \, , \qquad [9.4]$$

or

$$\lim_{N \to \infty} \int |R_N(x)|^2 \, dx = 0 \, . \qquad [9.5]$$

Because of [9.1] and [9.3] we have

$$\int dx |f(x) - \sum_{n=1}^{N} a_n u_n(x)|^2 = \int dx |f(x)|^2 - \sum_{n=1}^{N} a_n \int f^*(x) u_n(x) \, dx$$

$$- \sum_{n=1}^{N} a_n^* \int f(x) u_n^*(x) \, dx + \sum_{n=1}^{N} a_n^* a_n = \int |f|^2 \, dx - \sum_{n=1}^{N} a_n^* a_n \, . \qquad [9.6]$$

Because the left side of [9.6] is certainly not negative, Bessel's inequality follows:

$$\sum_{n=1}^{N} a_n^* a_n \leqslant \int |f|^2 \, dx, \quad \text{or} \quad \sum_{n=1}^{\infty} |a_n|^2 \leqslant \int |f|^2 dx \, . \qquad [9.7]$$

Bessel's inequality implies that [9.4] means the same as

$$\sum_{n=1}^{\infty} |a_n|^2 = \int |f|^2 dx \, . \qquad [9.8]$$

This equation is known as the completeness relation because it guarantees us that none of the u_n were left out, that is, that the set of u_n is complete.[3]

The generalization to two or more functions is simple with the substitution

$$f(x) \to c_1 f(x) + c_2 g(x),$$

$$a_n \to c_1 a_n + c_2 b_n, \qquad b_n = \int g(x) u_n^*(x) \, dx \, .$$

---

[3] Since we are dealing here with sets of infinitely many functions, we cannot establish completeness by counting, for example.

Substitution into [9.8] and comparison of the coefficients of $c_1^* c_2$ yields, for example,

$$\sum_{n=1}^{\infty} a_n^* b_n = \int f^*(x) g(x) \, \mathrm{d}x \ . \qquad [9.9]$$

By means of the δ-function, the completeness relation can also be written symbolically as

$$\sum_n u_n^*(x) u_n(x') = \delta(x - x') \ . \qquad [9.10]$$

This can be verified in the following way. If we multiply the above equation by $f^*(x')$ and integrate, then we obtain with [9.3]

$$\sum_n u_n^*(x) \int u_n(x') f^*(x') \, \mathrm{d}x' = f^*(x) \ . \qquad [9.11]$$

Multiplying this equation by $g(x)$ and integrating once again yields

$$\sum_n \int g(x) u_n^*(x) \, \mathrm{d}x \int u_n(x') f^*(x') \, \mathrm{d}x' = \int f^*(x) g(x) \, \mathrm{d}x \ , \qquad [9.12]$$

which is [9.9].

Remark: Equation [9.11] is in fact correct in practically all cases which arise in physics. However, since we have only required convergence in the mean for the series, we must integrate again for the sake of exactness. That is, we must go over to Eq. [9.12].

10. INITIAL-VALUE PROBLEM AND THE FUNDAMENTAL SOLUTION

We call that solution of the wave equation which describes a particle that was definitely at position x' at time $t = 0$,

$$\psi(x, 0) = \delta(x - x') \ , \qquad [10.1]$$

the fundamental solution $K(x, x', t)$.

We first consider a particle in a box. The general solution

for that case is (see Sec. 7)

$$\psi(x, t) = \sum_n c_n u_n(x) \exp\left[-\frac{i}{h} E_n t\right]. \qquad [10.2]$$

By using the completeness relation [9.10] we can also write down the fundamental solution immediately:

$$\psi(x, t) \equiv K(x, x', t) = \sum_n u_n(x) u_n^*(x') \exp\left[-\frac{i}{h} E_n t\right], \qquad [10.3]$$

or, for a three-dimensional region with arbitrary boundaries,

$$K(\boldsymbol{x}, \boldsymbol{x}', t) = \sum_n u_n(\boldsymbol{x}) u_n^*(\boldsymbol{x}') \exp\left[-\frac{i}{h} E_n t\right]. \qquad [10.4]$$

Whereas only a linear superposition of stationary states is possible as the wave function of a particle in a box (the boundary conditions must certainly be satisfied at all times), there is no such restriction for a particle in free space. In that case we have the continuous variable p instead of the discrete index n (see Sec. 8), and the wave function is

$$\psi(x, t) = \int \varphi(p) u(p, x) \exp\left[-\frac{i}{h} \frac{p^2}{2m} t\right] dp$$

$$= \frac{1}{\sqrt{2\pi h}} \int \varphi(p) \exp\left[\frac{i}{h}\left(px - \frac{p^2}{2m} t\right)\right] dp \qquad [10.5]$$

with the normalization

$$\int u^*(p, x) u(p', x)\, dx = \delta(p - p'). \qquad [10.6]$$

The completeness relation [9.10], which is now written in the form

$$\int u(p, x) u^*(p, x')\, dp = \delta(x - x') \qquad [10.7]$$

(note the symmetry between x and p!), again allows us to write down the fundamental solution immediately:

$$\psi(x, t) = \int u(p, x) u^*(p, x') \exp\left[-\frac{i}{h} \frac{p^2}{2m} t\right] dp, \qquad [10.8]$$

or, by comparison with [10.5],

$$\psi(x, t) \equiv K(x - x', t)$$
$$= \frac{1}{2\pi h} \int \exp\left\{\frac{i}{h}\left[p(x - x') - \frac{p^2}{2m}t\right]\right\} dp .$$ [10.9]

We can perform the integration:

$$K(x - x', t) = \frac{1}{2\pi h} \exp\left[\frac{i}{h} \cdot \frac{m}{2} \cdot \frac{(x - x')^2}{t}\right]$$
$$\times \int \exp\left[-\frac{i}{h}\left(p - \frac{(x - x')m}{t}\right)^2 \cdot \frac{t}{2m}\right] dp .$$

The substitution

$$p - \frac{(x - x')m}{t} = \sqrt{\frac{2mh}{t}} \cdot u$$

leads to a Fresnel integral:

$$\int \exp[iu^2] du = \sqrt{\pi} \cdot \exp\left[i\frac{\pi}{4}\right],$$
$$\int \exp[-iu^2] du = \sqrt{\pi} \cdot \exp\left[-i\frac{\pi}{4}\right] = \frac{\sqrt{\pi}}{\sqrt{i}} .$$ [10.10]

The solution $K(x - x', t)$ can now be written as

$$K(x - x', t) = \frac{1}{\sqrt{2\pi hi}} \cdot \sqrt{\frac{m}{t}} \cdot \exp\left[\frac{i}{h} \cdot \frac{m(x - x')^2}{2t}\right] .$$ [10.11]

Using [10.10] it also follows immediately that

$$\int K(x - x', t) dx = 1 .$$ [10.12]

Now, without using the Dirac function, we want to show that the $K(x - x', t)$ defined by [10.11] has the properties of the fundamental solution. Let $\xi \equiv x - x'$. We shall calculate

$$\lim_{t \to 0} \int_{\xi_1}^{\xi_2} K(\xi, t) d\xi .$$

With the substitution

$$\sqrt{\frac{m}{2ht}} \cdot \xi = u$$

we obtain

$$\int_{\xi_1}^{\xi_2} K(\xi, t)\, d\xi = \frac{\exp\left[-i(\pi/4)\right]}{\sqrt{2\pi h}} \sqrt{\frac{m}{t}}\, \sqrt{\frac{2ht}{m}} \int_{u_1}^{u_2} \exp[iu^2]\, du .$$

There are three cases to be considered:

$$
\begin{aligned}
a.\quad & \xi_1 > 0, \quad \xi_2 > 0: \quad & u_1 > 0 \quad & u_2 > 0 \\
& t \to 0: \quad & u_1 \to +\infty \quad & u_2 \to +\infty
\end{aligned}
\left.\begin{aligned} \\ \\ \\ \\ \end{aligned}\right\} \lim_{t \to 0} \int_{\xi_1}^{\xi_2} K\, d\xi = 0,
$$

$$
\begin{aligned}
b.\quad & \xi_1 < 0, \quad \xi_2 < 0: \quad & u_1 < 0 \quad & u_2 < 0 \\
& t \to 0: \quad & u_1 \to -\infty \quad & u_2 \to -\infty
\end{aligned}
$$

$$
\begin{aligned}
c.\quad & \xi_1 < 0, \quad \xi_2 > 0: \quad & u_1 < 0 \quad & u_2 > 0 \\
& t \to 0: \quad & u_1 \to -\infty \quad & u_2 \to +\infty
\end{aligned}
\left.\begin{aligned} \\ \\ \end{aligned}\right\} \lim_{t \to 0} \int_{\xi_1}^{\xi_2} K\, d\xi = 1.
$$

The formulas in [10.10] were used in evaluating the last case. The three cases a, b, and c taken together are equivalent to the equation

$$K(\xi, 0) = \delta(\xi) ,$$

and it remains to be shown that $K(\xi, t)$ satisfies the wave equation

$$\frac{\partial K}{\partial t} = \frac{ih}{2m} \cdot \frac{\partial^2 K}{\partial \xi^2} :$$

$$\frac{\partial K}{\partial t} = \frac{\sqrt{m}}{\sqrt{2\pi h i}}\left(-\frac{1}{2}\frac{1}{t\sqrt{t}} - \frac{1}{\sqrt{t}} \cdot \frac{i}{h} \cdot \frac{m\xi^2}{2t^2}\right) \exp\left[\frac{i}{h}\frac{m\xi^2}{2t}\right],$$

$$\frac{\partial K}{\partial \xi} = \sqrt{\frac{m}{2\pi h i}} \cdot \frac{1}{\sqrt{t}} \cdot \frac{im}{ht} \cdot \xi \cdot \exp\left[\frac{i}{h}\frac{m\xi^2}{2t}\right],$$

$$\frac{\partial^2 K}{\partial \xi^2} = \sqrt{\frac{m}{2\pi h i}} \frac{1}{\sqrt{t}}\left(\frac{im}{ht} - \frac{m^2}{h^2 t^2}\xi^2\right) \exp\left[\frac{i}{h}\frac{m\xi^2}{2t}\right] = \frac{2m}{ih} \cdot \frac{\partial K}{\partial t} .$$

Q.E.D.

It is worth noting that the fundamental solution [10.11] is well known from the theory of heat conduction. If we replace the mass m in the nonrelativistic wave equation by a purely imaginary quantity, then indeed we obtain the heat conduction equation, and the Gaussian distribution [10.11] is the familiar solution for a δ-function heat pulse at time $t = 0$.

Chapter 3. Particle in a Field of Force

11. THE HAMILTONIAN OPERATOR

By means of the Hamiltonian function

$$H = \sum_{k=1}^{3} \frac{p_k^2}{2m} + V(x_1, x_2, x_3, t) , \qquad [11.1]$$

the equations of motion of classical mechanics can be written in the canonical form

$$\left. \begin{aligned} \frac{dp_k}{dt} &= -\frac{\partial H}{\partial x_k} = -\frac{\partial V}{\partial x_k} \\ \frac{dx_k}{dt} &= +\frac{\partial H}{\partial p_k} = \frac{p_k}{m} \end{aligned} \right\} . \qquad [11.2]$$

We obtain the corresponding equations for a particle of charge e in a magnetic field H by introducing the vector potential A $(H = \nabla \times A)$:

$$H = \sum_{k=1}^{3} \frac{1}{2m} \left\{ p_k - \frac{e}{c} A_k(x) \right\}^2 + V(x_1, x_2, x_3) , \qquad [11.3]$$

$$\left. \begin{aligned} \frac{dp_k}{dt} &= -\frac{\partial H}{\partial x_k} = -\frac{\partial V}{\partial x_k} + \frac{1}{m} \sum_i \left(p_i - \frac{e}{c} A_i \right) \cdot \frac{e}{c} \frac{\partial A_i}{\partial x_k} \\ \frac{dx_k}{dt} &= +\frac{\partial H}{\partial p_k} = \frac{1}{m} \left(p_k - \frac{e}{c} A_k \right) \end{aligned} \right\} . \qquad [11.4]$$

Again, we use the operator formalism given in [2.4] as a translational key between classical mechanics and wave

38

mechanics:

$$p_k \to - ih \frac{\partial}{\partial x_k}, \qquad E \to ih \frac{\partial}{\partial t}. \qquad [11.5]$$

As a result of introducing the operators, the Hamiltonian function becomes the so-called Hamiltonian operator:

$$H(p_i, x_i) \to \underline{H}\left(\frac{\partial}{\partial x_i}, x_i\right). \qquad [11.6]$$

Generally, an operator $\underline{A}$ is understood to be a calculational prescription which associates a function u with another function, called $\underline{A}u$.[1]

In the force-free case ($V \equiv 0$), we can also write the wave equation [2.11] with the operators [11.5] and [11.6]:

$$\underline{H}\psi = \underline{E}\psi. \qquad [11.7]$$

This equation also gives the correct result for a particle in a field of force ($V \neq 0$). Written out, the equation then reads

$$-\frac{h^2}{2m} \nabla^2 \psi + V(x)\psi = ih \frac{\partial \psi}{\partial t}; \qquad [11.8]$$

it was first suggested in this form by Schrödinger.[2]

It so happens that Eq. [11.7] is also correct in more complicated cases, for example, in the case of a charged particle in a magnetic field. Nevertheless, an ambiguity sometimes arises when the classical Hamiltonian function is replaced by the Hamiltonian operator. Whereas the ordering of factors in a product is of no importance for classical quantities,

$$p_i x_k = x_k p_i,$$

with operators the ordering is very important. For ex-

---

[1] It is not necessary that the prescription be a differentiation. For example, the operator $\underline{x}$ is defined to mean multiplication by x.

[2] E. Schrödinger, *Ann. Physik* **81**, 109 (1926).

ample, later we shall find that

$$p_i x_k - x_k p_i = \frac{h}{i} \delta_{ik} .$$

Thus, it is possible for several different operators to be derived from a given Hamiltonian function.[3] Which of these different possibilities is correct can be decided only by experiment; that is, the expectation values derived from the solution of [11.7] must follow the classical trajectories.

12. HERMITIAN OPERATORS

The solutions of the wave equation [11.7] must satisfy the important condition that their normalization be constant:

$$\frac{\partial}{\partial t} \int \psi^* \psi \, \mathrm{d}^3 x = 0 . \qquad [12.1]$$

(This condition can easily be understood physically by remembering the interpretation of $\psi^* \psi$ as a probability density.)

Carrying out the differentiation in [12.1] and using the equations

$$-\frac{h}{i} \cdot \frac{\partial \psi}{\partial t} = \underline{H} \psi \qquad \text{and} \qquad +\frac{h}{i} \cdot \frac{\partial \psi^*}{\partial t} = (\underline{H} \psi)^* , \qquad [12.2]$$

we obtain

$$\int [\psi^*(\underline{H}\psi) - \psi(\underline{H}\psi)^*] \, \mathrm{d}^3 x = 0 . \qquad [12.3]$$

This is a condition on $\underline{H}$. Any operator which satisfies this relation is called *Hermitian*.

Since the superposition principle should be valid for the

---

[3] Cross terms, such as $p_i x_k$, which lead to ambiguities occur with non-Cartesian coordinates, for example.

waves, we must require

$$\left.\begin{array}{l} \underline{H}(c\psi) = c\underline{H}\psi \\ \underline{H}(\psi_1 + \psi_2) = \underline{H}\psi_1 + \underline{H}\psi_2 \end{array}\right\}. \qquad [12.4]$$

An operator with this property is called *linear*.

If we substitute

$$\psi = \psi_1 + \psi_2$$

in Eq. [12.3], we obtain

$$\int [\psi_1^*(\underline{H}\psi_2) - \psi_2(\underline{H}\psi_1)^*]\,\mathrm{d}^3x = 0 \qquad [12.5]$$

for a linear Hermitian operator $\underline{H}$. If $\underline{F}$ and $\underline{G}$ are linear Hermitian operators and a and b are real numbers, then we have

$$a\underline{F} + b\underline{G} \qquad \text{Hermitian}, \qquad [12.6]$$

$$\underline{F}\,\underline{G} + \underline{G}\,\underline{F} \qquad \text{Hermitian}, \qquad [12.7]$$

$$i(\underline{F}\,\underline{G} - \underline{G}\,\underline{F}) \qquad \text{Hermitian}, \qquad [12.8]$$

but $\underline{F}\,\underline{G}$ in general is not Hermitian.

Proof of [12.7]: From [12.5] we have

$$\int \{(\underline{H}_1\psi_1)^*\underline{H}_2\psi_2 - \psi_2(\underline{H}_2\underline{H}_1\psi_1)^*\}\,\mathrm{d}^3x = 0 , \qquad [a]$$

$$\int \{(\underline{H}_2\psi_2)^*\underline{H}_1\psi_1 - \psi_1(\underline{H}_1\underline{H}_2\psi_2)^*\}\,\mathrm{d}^3x = 0 ,$$

$$\int \{\underline{H}_2\psi_2(\underline{H}_1\psi_1)^* - \psi_1^*\underline{H}_1\underline{H}_2\psi_2\}\,\mathrm{d}^3x = 0 . \qquad [b]$$

$[a]-[b]$:

$$\int \{\psi_1^*\underline{H}_1\underline{H}_2\psi_2 - \psi_2(\underline{H}_2\underline{H}_1\psi_1)^*\}\,\mathrm{d}^3x = 0 . \qquad [c]$$

Substitute $\underline{H}_1 = \underline{F}$, $\underline{H}_2 = \underline{G}$ in [c]; then substitute $\underline{H}_1 = \underline{G}$,

$\underline{H}_2 = \underline{F}$ and add:

$$\int \{\psi_1^*(\underline{F}\,\underline{G} + \underline{G}\,\underline{F})\psi_2 - \psi_2[(\underline{G}\,\underline{F} + \underline{F}\,\underline{G})\psi_1]^*\}\,\mathrm{d}^3x = 0 \ . \qquad \text{Q.E.D.}$$

Examples of Hermitian operators: The operator

$$\underline{p} = -ih(\partial/\partial x)$$

is Hermitian, as can be seen by partial integration:

$$\int \psi^* \frac{h}{i} \frac{\partial \psi}{\partial x}\,\mathrm{d}x = -\frac{h}{i}\int \frac{\partial \psi^*}{\partial x}\,\psi\,\mathrm{d}x = \int \left(\frac{h}{i}\frac{\partial \psi}{\partial x}\right)^*\psi\,\mathrm{d}x \ .$$

On the other hand, $i\underline{p}$ is not Hermitian. The operator $\underline{p}^2 = -h^2(\partial^2/\partial x^2)$ is also Hermitian because

$$\int \left(\psi^* \frac{\partial^2 \psi}{\partial x^2} - \frac{\partial^2 \psi^*}{\partial x^2}\,\psi\right)\mathrm{d}x = 0 \qquad \text{(Green's formula)} \ .$$

Any operator which prescribes multiplication by a real function of x is clearly Hermitian. Thus, the potential $\underline{V}(x)$ is Hermitian, and this means that our Hamiltonian operator,

$$\underline{H}\psi = \left(\frac{\underline{p}^2}{2m} + \underline{V}(\boldsymbol{x})\right)\psi = -\frac{h^2}{2m}\nabla^2\psi + V(\boldsymbol{x})\psi \ , \qquad [12.9]$$

is also Hermitian.

$\underline{H}$ remains Hermitian even in the case of a magnetic field. When we write the Hamiltonian operator corresponding to [11.3] we need only pay attention to the ordering of the factors: instead of writing

$$\left(p_k - \frac{e}{c}A_k\right)^2 = p_k^2 - 2\frac{e}{c}p_kA_k + \frac{e^2}{c^2}A_k^2 \ ,$$

we must write

$$\left(p_k - \frac{e}{c}A_k\right)^2 = p_k^2 - \frac{e}{c}p_kA_k - \frac{e}{c}A_kp_k + \frac{e^2}{c^2}A_k^2 \ .$$

In order to prove the Hermiticity of the Hamiltonian operator that is derived in this manner from [11.3], all we need show is that

$$p_k A_k + A_k p_k$$

is Hermitian (special case of formula [12.7]). This follows by partial integration:

$$\int \psi^* \left\{ \frac{h}{i} \frac{\partial}{\partial x_k} (A_k \psi) + A_k \frac{h}{i} \frac{\partial \psi}{\partial x_k} \right\} d^3x$$

$$= -\int \left\{ \frac{h}{i} \frac{\partial \psi^*}{\partial x_k} (A_k \psi) + \frac{h}{i} \frac{\partial}{\partial x_k} (\psi^* A_k) \psi \right\} d^3x$$

$$= \int \psi \left\{ \frac{h}{i} \frac{\partial \psi}{\partial x_k} A_k + \frac{h}{i} \frac{\partial}{\partial x_k} (\psi A_k) \right\}^* d^3x .$$

13. EXPECTATION VALUES AND THE CLASSICAL EQUATION OF MOTION. COMMUTATION RELATIONS (COMMUTATORS)

So far we have learned to calculate the probability distribution for observable quantities by using wave functions. If classical mechanics is to be contained in wave mechanics as a limiting case, then the expectation values of these quantities must obey the classical equations of motion. We now demonstrate that this is the case.

First we calculate the time derivative of $\langle x_k \rangle$ (see [3.5], as well as [12.2] and [12.5]):

$$\langle x_k \rangle = \int \psi^* x_k \psi \, d^3x ;$$

$$\frac{d}{dt} \langle x_k \rangle = \frac{i}{h} \int [(H\psi)^* x_k \psi - \psi^* x_k (H\psi)] \, d^3x$$

$$= \frac{i}{h} \int [\psi^* (H x_k \psi) - \psi^* x_k H \psi] \, d^3x$$

$$= \frac{i}{h} \int \psi^* (H x_k - x_k H) \psi \, d^3x ,$$

$$\frac{d}{dt} \langle x_k \rangle = \frac{i}{h} \langle H x_k - x_k H \rangle . \tag{13.1}$$

In a similar way, we can derive

$$\frac{d}{dt}\langle \underline{p}_k \rangle = \frac{i}{h}\langle \underline{H}\,\underline{p}_k - \underline{p}_k\,\underline{H} \rangle .$$ [13.2]

Now, we must evaluate the right-hand sides of [13.1] and [13.2]. In general, an expression of the form

$$\underline{F}_1\underline{F}_2 - \underline{F}_2\underline{F}_1 \equiv [\underline{F}_1, \underline{F}_2]$$ [13.3]

is called a *commutator*. The relations

$$[\underline{F}_1\underline{F}_2, \underline{F}_3] \equiv \underline{F}_1[\underline{F}_2, \underline{F}_3] + [\underline{F}_1, \underline{F}_3]\underline{F}_2$$ [13.4]

and

$$[c_1\underline{F}_1 + c_2\underline{F}_2, \underline{F}_3] \equiv c_1[\underline{F}_1, \underline{F}_3] + c_2[\underline{F}_2, \underline{F}_3]$$ [13.5]

follow immediately from the definition [13.3].

In agreement with the formulas for Fourier transformation, [4.3] and [4.4], we now define the operators analogous to [12.9] associated with the momenta and coordinates:

$$\underline{p}_k\psi(\boldsymbol{x}) = \frac{h}{i}\cdot\frac{\partial}{\partial x_k}\psi , \qquad \underline{x}_k\psi(\boldsymbol{x}) = x_k\psi ,$$ [13.6]

$$\underline{p}_k\varphi(\boldsymbol{p}) = p_k\varphi , \qquad \underline{x}_k\varphi(\boldsymbol{p}) = -\frac{h}{i}\cdot\frac{\partial}{\partial p_k}\varphi .$$ [13.7]

With these definitions we can immediately derive the basic commutation relations:

$$\left.\begin{array}{l} \underline{p}_k\underline{x}_l - \underline{x}_l\underline{p}_k = \delta_{lk}\cdot\dfrac{h}{i} \\[2mm] \underline{p}_k\underline{p}_l - \underline{p}_l\underline{p}_k = 0 \\[2mm] \underline{x}_k\underline{x}_l - \underline{x}_l\underline{x}_k = 0 \end{array}\right\} .$$ [13.8]

For example:

$$\underline{p}_k\underline{x}_l\psi = \frac{h}{i}\cdot\frac{\partial}{\partial x_k}(x_l\psi) , \qquad \underline{x}_l\underline{p}_k\psi = x_l\frac{h}{i}\cdot\frac{\partial}{\partial x_k}\psi ;$$

$$(\underline{p}_k\underline{x}_l - \underline{x}_l\underline{p}_k)\psi = \frac{h}{i}\left(\frac{\partial}{\partial x_k}(x_l\psi) - x_l\frac{\partial}{\partial x_k}\psi\right) = \delta_{kl}\frac{h}{i}\psi .$$

Using [13.4] and [13.8] we obtain

$$\sum_i (\underline{p}_i^2 \underline{x}_k - \underline{x}_k \underline{p}_i^2) = \sum_i \{\underline{p}_i(\underline{p}_i \underline{x}_k - \underline{x}_k \underline{p}_i) + (\underline{p}_i \underline{x}_k - \underline{x}_k \underline{p}_i)\underline{p}_i\} =$$
$$= 2\frac{h}{i}\underline{p}_k = \frac{h}{i}\frac{\partial}{\partial \underline{p}_k}\sum_i \underline{p}_i^2 .$$

Here we have introduced the new operation of differentiating a function of operators with respect to an operator. We define that this differentiation is to be performed in exactly the usual way in which a function is differentiated with respect to an ordinary variable.

By repeated application of this procedure we obtain

$$\underline{F}(\underline{p})\underline{x}_k - \underline{x}_k \underline{F}(\underline{p}) = \frac{h}{i}\frac{\partial \underline{F}}{\partial \underline{p}_k} \qquad [13.9]$$

for an operator $\underline{F}$ which is given as a polynomial in $\underline{p}_k$, or, in the limiting case, as a convergent power series in $\underline{p}_k$ (that is, as an analytic function of $\underline{p}_k$). Of course, the relation

$$\underline{F}(\underline{p})\underline{p}_k - \underline{p}_k \underline{F}(\underline{p}) = 0 \qquad [13.10]$$

also holds. In an analogous way, using [4.3] and [4.4], we obtain

$$\underline{G}(\underline{x})\underline{p}_k - \underline{p}_k \underline{G}(\underline{x}) = -\frac{h}{i}\frac{\partial \underline{G}}{\partial \underline{x}_k} , \qquad [13.11]$$

$$\underline{G}(\underline{x})\underline{x}_k - \underline{x}_k \underline{G}(\underline{x}) = 0 . \qquad [13.12]$$

Summation of the last four equations yields

$$\underline{H}\,\underline{p}_k - \underline{p}_k\underline{H} = -\frac{h}{i}\cdot\frac{\partial \underline{H}}{\partial \underline{x}_k} , \qquad [13.13]$$

$$\underline{H}\,\underline{x}_k - \underline{x}_k\underline{H} = \frac{h}{i}\cdot\frac{\partial \underline{H}}{\partial \underline{p}_k} \qquad [13.14]$$

for an operator

$$\underline{H}(\underline{p}, \underline{x}) = \underline{F}(\underline{p}) + \underline{G}(\underline{x}) . \qquad [13.15]$$

Thus, the result applies to the Hamiltonian operator with a potential. It is easy to show that [13.13] and [13.14] are also valid for an operator

$$\underline{H}(\underline{p}, \underline{x}) = \underline{F}(\underline{p}) + \underline{G}(\underline{x}) + \sum_k \{\underline{A}_k(\underline{x})\underline{p}_k + \underline{p}_k \underline{A}_k(\underline{x})\} \quad [13.16]$$

(Hamiltonian operator with a magnetic field). If we substitute [13.13] and [13.14] into [13.1] and [13.2], we obtain the classical canonical equations of motion for the average values of x_k and p_k:

$$\frac{d}{dt} \langle \underline{x}_k \rangle = \left\langle \frac{\partial \underline{H}}{\partial \underline{p}_k} \right\rangle, \qquad \frac{d}{dt} \langle \underline{p}_k \rangle = - \left\langle \frac{\partial \underline{H}}{\partial \underline{x}_k} \right\rangle. \qquad [13.17]$$

(That is not purely by chance; rather, the theory was so constructed that [13.17] would result. It will always be required that a reasonable theory contain classical mechanics as a limiting case.)

As a generalization we proceed to consider the case in which an operator $\underline{F}$ depends on time explicitly (for example, time-dependent forces). From

$$\langle \underline{F} \rangle = \int \psi^*(\underline{F}\psi) \, d^3x$$

and the wave equations [12.2] follows

$$\frac{d}{dt} \langle \underline{F} \rangle = \frac{i}{h} \int \{(\underline{H}\psi)^*(\underline{F}\psi) - \psi^*\underline{F}(\underline{H}\psi)\} \, d^3x + \int \psi^* \frac{\partial \underline{F}}{\partial t} \psi \, d^3x$$

$$= \int \psi^* \left\{ \frac{i}{h}(\underline{H}\,\underline{F} - \underline{F}\,\underline{H}) + \frac{\partial \underline{F}}{\partial t} \right\} \psi \, d^3x , \qquad \left(\text{see } [12.5]\right)$$

$$\frac{d}{dt} \langle \underline{F} \rangle = \left\langle \frac{i}{h}[\underline{H}, \underline{F}] + \frac{\partial \underline{F}}{\partial t} \right\rangle. \qquad [13.18]$$

For $\underline{F} = \underline{H}$ we obtain

$$\frac{d}{dt} \langle \underline{H} \rangle = \left\langle \frac{\partial \underline{H}}{\partial t} \right\rangle. \qquad [13.19]$$

If $\delta \underline{H}/\partial t = 0$, it follows that $\langle \underline{H} \rangle = $ constant; this is a statement of the law of conservation of energy.

As an example, let us specialize to the case of a magnetic field:

$$\underline{H} = \frac{1}{2m} \sum_k \left(\underline{p}_k - \frac{e}{c} \underline{A}_k(\boldsymbol{x}, t) \right)^2 + \underline{V}(\boldsymbol{x}, t)$$

$$= \frac{1}{2m} \sum_k \left(\underline{p}_k^2 - \frac{e}{c} (\underline{p}_k \underline{A}_k + \underline{A}_k \underline{p}_k) + \frac{e^2}{c^2} \underline{A}_k^2 \right) + \underline{V} . \qquad [13.20]$$

Using [13.17] we obtain

$$\frac{d}{dt} \langle \underline{x}_k \rangle = \frac{1}{m} \left\langle \underline{p}_k - \frac{e}{c} \underline{A}_k \right\rangle , \qquad [13.21]$$

$$\frac{d}{dt} \langle \underline{p}_k \rangle = \left\langle \frac{1}{2m} \frac{e}{c} \sum_i \left\{ \left(\underline{p}_i \frac{\partial \underline{A}_i}{\partial \underline{x}_k} + \frac{\partial \underline{A}_i}{\partial \underline{x}_k} \underline{p}_i \right) \right. \right.$$

$$\left. - \frac{e}{c} \left(\frac{\partial \underline{A}_i}{\partial \underline{x}_k} \underline{A}_i + \underline{A}_i \frac{\partial \underline{A}_i}{\partial \underline{x}_k} \right) \right\} - \frac{\partial \underline{V}}{\partial \underline{x}_k} \right\rangle$$

$$= \left\langle \frac{1}{2m} \frac{e}{c} \sum_i \left\{ \left(\underline{p}_i - \frac{e}{c} \underline{A}_i \right) \frac{\partial \underline{A}_i}{\partial \underline{x}_k} \right. \right.$$

$$\left. + \frac{\partial \underline{A}_i}{\partial \underline{x}_k} \left(\underline{p}_i - \frac{e}{c} \underline{A}_i \right) \right\} - \frac{\partial \underline{V}}{\partial \underline{x}_k} \right\rangle . \qquad [13.22]$$

Note the symmetrization, which is necessary for preserving Hermiticity.

By not using the canonical form we can also write the equation of motion more simply:

$$\langle m \underline{\ddot{x}}_k \rangle = \left\langle \underline{\dot{p}}_k - \frac{e}{c} \underline{\dot{A}}_k (\boldsymbol{x}) \right\rangle , \qquad \langle m \underline{\ddot{x}}_k \rangle = \langle \underline{K}_k \rangle , \qquad [13.23]$$

where the definition

$$\underline{\dot{F}} = \frac{i}{h} [\underline{H}, \underline{F}] + \frac{\partial \underline{F}}{\partial t}$$

is used and

$$\underline{K}_k = - \frac{\partial \underline{V}}{\partial \underline{x}_k} + e \underline{E}_k + \frac{e}{2c} \sum_i (\underline{H}_{ki} \underline{\dot{x}}_i + \underline{\dot{x}}_i \underline{H}_{ki}) , \qquad [13.24]$$

with a magnetic field (written as an antisymmetric tensor)

$$\underline{H}_{ki} = \frac{\partial \underline{A}_i}{\partial \underline{x}_k} - \frac{\partial \underline{A}_k}{\partial \underline{x}_i} , \qquad [13.25]$$

and with an induced electric field

$$\underline{E}_k = -\frac{1}{c}\frac{\partial \underline{A}_k}{\partial t} - \frac{\partial \underline{V}}{\partial \underline{x}_k} . \qquad [13.26]$$

Equation [13.22] can be easily derived from [13.23] and [13.24].

We also have a continuity equation if there is a magnetic field (see [7.11]):

$$\frac{\partial}{\partial t}(\psi^*\psi) + \mathrm{div}\,\boldsymbol{i} = 0 , \qquad [13.27]$$

where the density $\boldsymbol{i}$ of the probability current is given by

$$i_k = \frac{1}{2m}\left\{\psi^*\left(\underline{p}_k - \frac{e}{c}\underline{A}_k\right)\psi - \psi\left(\underline{p}_k + \frac{e}{c}\underline{A}_k\right)\psi^*\right\} \qquad [13.28]$$

or

$$i_k = \frac{h}{2mi}\left(\psi^*\frac{\partial \psi}{\partial x_k} - \psi\frac{\partial \psi^*}{\partial x_k}\right) - \frac{e}{mc}\underline{A}_k\,\psi^*\psi . \qquad [13.29]$$

Gauge invariance

As we know from electrodynamics,[4] a gradient can be added to the vector potential $\boldsymbol{A}$ without changing the magnetic field:

$$A_k \rightarrow A_k + \frac{\partial f(\boldsymbol{x}, t)}{\partial x_k} . \qquad [13.30]$$

However, because of [13.26], an extra term must then be added to the potential,

$$V \rightarrow V - \frac{e}{c}\frac{\partial f}{\partial t}; \qquad [13.31]$$

[4] See W. PAULI, *Lectures in Physics: Electrodynamics* (M.I.T. Press, Cambridge, Mass., 1972).

this means that the form of the Hamiltonian operator is changed. However, from the wave equation [12.2], it follows that the only change in ψ is that it is multiplied by a phase factor:

$$\psi \to \psi \cdot \exp\left[\frac{i}{h}\cdot\frac{e}{c}\cdot f\right]. \qquad [13.32]$$

The group of transformations defined by [13.30], [13.31], and [13.32] is known as the *gauge group*. Correspondingly, those quantities which do not change under these transformations are said to be *gauge invariant*. Examples of such quantities are the probability density $\psi^*\psi$ and the probability current i.

As a further example we consider the virial theorem.[5] The quantum theoretical expression

$$\left\langle \frac{m}{2}\frac{d^2}{dt^2}(x_i x_k) \right\rangle$$

$$= \left\langle \frac{1}{2}(K_i x_k + x_i K_k) + \frac{m}{2}(\dot{x}_i\dot{x}_k + \dot{x}_i\dot{x}_k) \right\rangle \qquad [13.33]$$

follows from the relations

$$\frac{d}{dt}(x_i x_k) = \dot{x}_i x_k + x_i \dot{x}_k \qquad [13.34]$$

and

$$\frac{d^2}{dt^2}(x_i x_k) = \ddot{x}_i x_k + x_i \ddot{x}_k + \dot{x}_i\dot{x}_k + \dot{x}_i\dot{x}_k; \qquad [13.35]$$

for $i = k$, the expression reduces to

$$\left\langle \frac{m}{2}\frac{d^2}{dt^2}(x_k^2) \right\rangle = \langle K_k x_k + m\dot{x}_k^2 \rangle. \qquad [13.36]$$

Formulas [13.33] and [13.36] are also valid classically; but the symmetrized form is required by quantum mechanics for K_k (Eq. [13.24]).

---

[5] See W. PAULI, *Lectures in Physics: Thermodynamics* and *Statistical Mechanics* (M.I.T. Press, Cambridge, Mass., 1972).

From

$$\delta x_i = x_i - \langle x_i \rangle \qquad [13.37]$$

it follows that

$$\langle (\delta x_i)^2 \rangle = \langle x_i^2 \rangle - (\langle x_i \rangle)^2 \qquad [13.38]$$

and

$$\langle \delta x_i \, \delta x_k \rangle = \langle x_i x_k \rangle - \langle x_i \rangle \langle x_k \rangle , \qquad [13.39]$$

as we know. Using these relations, we can derive the more general expressions

$$\left\langle \frac{m}{2} \frac{d^2}{dt^2} (\delta x_i \, \delta x_k) \right\rangle = \left\langle \frac{1}{2} (K_i \, \delta x_k + \delta x_i K_k) \right.$$

$$\left. + \frac{m}{2} (\delta \dot{x}_i \delta \dot{x}_k + \delta \dot{x}_i \, \delta \dot{x}_k) \right\rangle \qquad [13.40]$$

and

$$\left\langle \frac{m}{2} \frac{d^2}{dt^2} (\delta x_k)^2 \right\rangle = \langle K_k \, \delta x_k + m(\delta \dot{x}_k)^2 \rangle . \qquad [13.41]$$

In the force-free case ($A_k = V = 0$, $m\dot{x}_k = p_k$), we again have the formula analogous to [3.20]:

$$\left\langle \frac{m^2}{2} \frac{d^2}{dt^2} (\delta x_k)^2 \right\rangle = \langle (\delta p_k)^2 \rangle . \qquad [13.42]$$

Chapter 4. More than One Particle

14. MORE THAN ONE PARTICLE

When there is more than one particle, a new feature of the problem is the interaction between particles. According to the theory of relativity, one should take into account the finite speed of propagation c of the action of these forces. In relativistic quantum theory the finite speed of propagation will be taken into account. In our nonrelativistic approximation we set $c = \infty$ for the sake of simplicity; that is, the action of a force propagates instantaneously to all particles. For this reason, we are able to use a single time coordinate t, which means that time is not considered on the same footing as the $3N$ space coordinates. We shall assume the number of particles, N, to be constant; thus, no radiative dissociation processes will be considered.[1]

From here on we designate the space coordinates by $q_1, \ldots, q_f$ $(f = 3N)$:

$$\{x_1^{(1)}, x_2^{(1)}, x_3^{(1)}, x_1^{(2)}, x_2^{(2)}, \ldots, x_3^{(N)}\} = \{q_1, q_2, q_3, q_4, q_5, \ldots, q_f\} \,.$$

Similarly, we designate the momentum coordinates by

$$\{p_1, \ldots, p_f\}$$

and define

$$\mathrm{d}^f q = \mathrm{d}q_1 \cdot \mathrm{d}q_2 \cdot \ldots \cdot \mathrm{d}q_f \,, \qquad \mathrm{d}^f p = \mathrm{d}p_1 \cdot \mathrm{d}p_2 \cdot \ldots \cdot \mathrm{d}p_f \,.$$

---

[1] See W. Pauli, *Lectures in Physics: Selected Topics in Field Quantization* (M.I.T. Press, Cambridge, Mass., 1972).

Further, we introduce two generalized wave functions $\psi(q_1, ..., q_f, t)$ and $\varphi(p_1, ..., p_f, t)$ which are defined such that

$$W(q_1, ..., q_f, t)\,\mathrm{d}'q = \psi^*\psi\,\mathrm{d}'q ,\qquad [14.1]$$

$$W(p_1, ..., p_f, t)\,\mathrm{d}'p = \varphi^*\varphi\,\mathrm{d}'p .\qquad [14.2]$$

These are the probabilities for finding the coordinates and momenta of the nth particle between q_i and $q_i + \mathrm{d}q_i$, and p_i and $p_i + \mathrm{d}p_i$, respectively ($i = 3n-2$, $3n-1$, $3n$; $n = 1, ..., N$).

In analogy to [4.3] and [4.4], we have the following relations:

$$\psi(q_1, ..., q_f, t) = \frac{1}{(2\pi h)^{f/2}} \int \varphi(p_1, ..., p_f, t)$$
$$\times \exp\left[+\frac{i}{h}(p_1 q_1 + p_2 q_2 + ... + p_f q_f)\right] \mathrm{d}'p ,\qquad [14.3]$$

$$\varphi(p_1, ..., p_f, t) = \frac{1}{(2\pi h)^{f/2}} \int \psi(q_1, ..., q_f, t)$$
$$\times \exp\left[-\frac{i}{h}(p_1 q_1 + p_2 q_2 + ... + p_f q_f)\right] \mathrm{d}'q .\qquad [14.4]$$

The wave function ψ is to satisfy the wave equation,

$$-\frac{h}{i}\frac{\partial\psi}{\partial t} = \underline{H}\psi ,\qquad [14.5]$$

where again $\underline{H}$ is a linear, Hermitian operator:

$$\int \psi_1^* \underline{H}\psi_2\,\mathrm{d}'q = \int \psi_2 (\underline{H}\psi_1)^*\,\mathrm{d}'q .\qquad [14.6]$$

Generalized to the $3N$ coordinates, the relations

$$\Delta p_k \cdot \Delta q_k \geqslant \frac{h}{2} ,\qquad [14.7]$$

$$p_k\psi \to \frac{h}{i}\cdot\frac{\partial}{\partial q_k}\psi , \qquad \text{and} \qquad q_k\varphi \to -\frac{h}{i}\cdot\frac{\partial}{\partial p_k}\varphi \qquad [14.8]$$

are also valid in this case.

Now we consider the special case in which there is no

interaction between the particles, which means that the Hamiltonian operator reduces to a sum of independent terms:

$$\underline{H} = \underline{H}^{(1)} + \underline{H}^{(2)} + \dots + \underline{H}^{(N)}. \qquad [14.9]$$

Here $\underline{H}^{(i)}$ is only to operate on the ith particle. If

$$\psi^{(1)}(q_1, q_2, q_3), \dots, \psi^{(N)}(q_{N-2}, q_{N-1}, q_N) \qquad [14.10]$$

are solutions of the wave equations

$$-\frac{h}{i}\frac{\partial \psi^{(a)}}{\partial t} = \underline{H}^{(a)}\psi^{(a)} \qquad (a = 1, 2, \dots, N) \qquad [14.11]$$

of the isolated systems, then

$$\psi = \psi^{(1)}\psi^{(2)} \dots \psi^{(N)} \qquad [14.12]$$

is a particular solution of

$$-\frac{h}{i}\frac{\partial \psi}{\partial t} = \underline{H}\psi = [\underline{H}^{(1)} + \underline{H}^{(2)} + \dots + \underline{H}^{(N)}]\psi. \qquad [14.13]$$

The general solution is a linear combination of products [14.12].

Thus, decomposition of the Hamiltonian operator into independent terms according to [14.9] corresponds to a reduction of the wave function to a product [14.12] of independent factors, in agreement with the fact that the probability reduces to a product when the particles are statistically independent:

$$\left.\begin{array}{l} W(q_1, \dots, q_f, t) = W_1(\boldsymbol{x}^{(1)}, t) \dots W_N(\boldsymbol{x}^{(N)}, t) \\ W(p_1, \dots, p_f, t) = W_1(\boldsymbol{p}^{(1)}, t) \dots W_N(\boldsymbol{p}^{(N)}, t) \end{array}\right\}. \qquad [14.14]$$

Therefore, we know the Hamiltonian operator for uncoupled particles in external fields of force (see [11.3] and [13.20]):

$$\underline{H}_0 = \sum_{a=1}^{N} \underline{H}^{(a)}; \qquad [14.15]$$

$$\underline{H}^{(a)} = -\frac{h^2}{2m^{(a)}}\sum_{k=1}^{3}\left(\frac{\partial}{\partial x_k^{(a)}} - \frac{i}{h}\frac{\epsilon^{(a)}}{c}A_k^{(a)}(\boldsymbol{x}^{(a)})\right)^2 + V^{(a)}(\boldsymbol{x}^{(a)}). \qquad [14.16]$$

If the forces between the particles can be derived from a potential $V(q)$, for example, in the case of a Coulomb interaction,

$$V(q_1, ..., q_f) = \sum_{a<b} \frac{e_a e_b}{r_{ab}}, \qquad [14.17]$$

where $r_{ab} = |x^{(a)} - x^{(b)}|$, then it is appropriate to write

$$\underline{H} = \underline{H}_0 + \underline{V}(q_1, ..., q_f) . \qquad [14.18]$$

The commutation relations [13.8], as well as formulas [13.13] and [13.14], are also valid in the case of more than one particle; the only modification necessary is that the indices i and k run from 1 through f.

Chapter 5. Eigenvalue Problems.
Functions of Mathematical Physics

We have already solved a particular eigenvalue problem in Section 7, the problem of a particle in a box. Now we have the necessary tools available for treating the more general eigenvalue problems of particles in more complicated fields of force.

Since we seek stationary solutions (standing waves) in eigenvalue problems, we use

$$\psi(x, t) = u(x) \exp\left[-\frac{i}{h} Et\right]$$

as a trial solution, in which case the wave equation [11.7] or [11.8] leads to the time-independent Schrödinger equation [1]

$$\underline{H}u = Eu .\qquad\qquad [I]$$

We now solve this equation in some simple cases.

15. THE LINEAR HARMONIC OSCILLATOR. HERMITE POLYNOMIALS

For the linear harmonic oscillator we have the well-known relations

$$H = \frac{p^2}{2m} + \frac{m}{2}\,\omega_0^2 q^2 = E ,\qquad\qquad [15.1]$$

$$\left.\begin{array}{l} p = -\dfrac{\partial H}{\partial q} = -m\omega_0^2 q \\[2mm] \dot{q} = \dfrac{\partial H}{\partial p} = \dfrac{p}{m} \end{array}\right\} \rightarrow \ddot{q} + \omega_0^2 q = 0 .\qquad [15.2]$$

---

[1] From a historical standpoint, it should be noted that the time-independent Schrodinger equation was found before the time-dependent one: E. SCHRODINGER, *Ann. Physik* **79**, 361 (1926).

From [15.1], corresponding to [I], we obtain the time-independent Schrödinger equation for the one-dimensional harmonic oscillator:

$$-\frac{h^2}{2m} \cdot \frac{d^2 u}{dq^2} + \frac{m}{2} \omega_0^2 q^2 u = Eu .$$ [15.3]

With the coordinate transformation

$$x = \sqrt{\frac{m\omega_0}{h}} \cdot q , \qquad p_x = \sqrt{\frac{1}{m\omega_0 h}} \cdot p \sim \frac{1}{i} \cdot \frac{\partial}{\partial x} ,$$ [15.4]

and with

$$\lambda \equiv \frac{2E}{h\omega_0} ,$$ [15.5]

[15.1] and [15.3] simplify to

$$H = \tfrac{1}{2} (p_x^2 + x^2) h\omega_0 ,$$ [15.6]

$$-\frac{d^2 u}{dx^2} + x^2 u = \lambda u ,$$ [15.7]

and the commutation relations [13.8] become simply

$$\underline{p}_x \underline{x} - \underline{x} \, \underline{p}_x = \frac{1}{i} .$$ [15.8]

With the substitution

$$u = y \exp\left[-\frac{x^2}{2}\right] ,$$ [15.9]

[15.7] becomes the differential equation

$$y'' - 2xy' + (\lambda - 1)y = 0 ,$$ [15.10]

which is solved by the Hermite polynomials. The Hermite polynomial of degree n is given by

$$H_n(x) = (-1)^n \cdot \exp[x^2] \cdot \frac{d^n}{dx^n} \exp[-x^2] .$$ [15.11]

By defining

$$\chi = \frac{d^n}{dx^n} \exp[-x^2] , \qquad [15.12]$$

we obtain with the polynomial rule of differentiation

$$\frac{d\chi}{dx} = \frac{d^n}{dx^n}(-2x \exp[-x^2])$$

$$= -2x \frac{d^n}{dx^n} \exp[-x^2] - 2n \frac{d^{n-1}}{dx^{n-1}} \exp[-x^2] , \qquad [15.13]$$

$$\frac{d^2\chi}{dx^2} = -2x \frac{d\chi}{dx} - 2(n+1)\chi , \qquad [15.14]$$

$$\chi'' + 2x\chi' + 2(n+1)\chi = 0 . \qquad [15.15]$$

From

$$y = \exp[x^2]\chi , \qquad \chi = \exp[-x^2]y ,$$

$$\chi' = \exp[-x^2](y' - 2xy) ,$$

$$\chi'' = \exp[-x^2]\{y'' - 4xy' + (4x^2 - 2)y\} \qquad [15.16]$$

follows

$$y'' - 2xy' + 2ny = 0 , \qquad [15.17]$$

and this equation is identical with [15.10] if we choose

$$\lambda = 2n + 1 . \qquad [15.18]$$

With this result, the solution of Eq. [15.7] can be written as

$$h_n(x) = c_n \exp\left[-\frac{x^2}{2}\right] H_n(x) , \qquad [15.19]$$

with c_n a constant and the eigenvalues (see [15.5])

$$E_n = h\omega_0(n + \tfrac{1}{2}) . \qquad [15.20]$$

Even in the case $n = 0$ there is a nonzero energy $E_0 = h\omega_0/2$ (*zero-point energy*). Now, we show that the eigenfunctions $h_n(x)$ are orthogonal; that is, they satisfy [7.12]. According to [15.7] and [15.18], we have

$$h_n'' + (2n + 1 - x^2)h_n = 0 ,$$

$$h_m'' + (2m + 1 - x^2)h_m = 0 .$$

Multiply the first equation by h_m and the second by $-h_n$. By adding the resulting equations and integrating, and using

$$h_m h_n'' - h_n h_m'' = (h_m h_n' - h_n h_m')',$$

we obtain

$$2(m-n)\int h_n h_m \, \mathrm{d}x = (h_m h_n' - h_n h_m')\Big|_{-\infty}^{+\infty} = 0, \qquad [15.21]$$

which for $n \neq m$, shows the validity of the orthogonality relations.

Before we can normalize the solution, we must derive a few more properties of the Hermite polynomials. The Hermite polynomials [15.11] are alternately even and odd:

$$\left.\begin{aligned}
H_0(x) &= 1, & H_1(x) &= 2x \\
H_2(x) &= 4x^2 - 2, & H_3(x) &= 8x^3 - 12x \\
H_4(x) &= 16x^4 - 48x^2 + 12, & \ldots\ldots\ldots
\end{aligned}\right\} \qquad [15.22]$$

Their generating function is

$$f(x, t) = \sum_{n=0}^{\infty} \frac{H_n(x)}{n!} t^n = \exp[x^2] \sum_{n=0}^{\infty} \frac{(-t)^n}{n!} \frac{\mathrm{d}^n}{\mathrm{d}x^n} \exp[-x^2]$$

$$= \exp[x^2]\exp[-(x-t)^2] = \exp[-t^2 + 2tx]. \quad [15.23]$$

From this,

$$\frac{\partial f}{\partial x} = 2tf \qquad [15.24]$$

follows immediately, and, by comparing coefficients, we obtain

$$H_n'(x) = 2n H_{n-1}(x). \qquad [15.25]$$

From

$$\frac{\partial f}{\partial t} + 2(t-x)f = 0 \qquad [15.26]$$

we obtain

$$x H_n(x) = \tfrac{1}{2} \cdot H_{n+1}(x) + n \cdot H_{n-1}(x) \qquad [15.27]$$

analogously. Of course, these relations, [15.25] and [15.27], also could have been derived directly from the definition [15.11]. From [15.25] follows

$$\frac{d^n}{dx^n} H_n(x) = 2n \cdot 2(n-1) \cdot \ldots \cdot 2 \cdot 1 \cdot H_0(x) = 2^n n! \qquad [15.28]$$

for the nth derivative. Now we can determine the constant c_n in [15.19] by normalizing

$$\int h_n^2(x)\, dx = c_n^2 \int \exp[-x^2] H_n^2(x) dx = 1. \qquad [15.29]$$

With [15.11] and [15.28] we find

$$c_n^{-2} = (-1)^n \int H_n(x) \frac{d^n}{dx^n} \exp[-x^2] dx$$

$$= \int \frac{d^n H_n(x)}{dx^n} \exp[-x^2]\, dx = 2^n n! \sqrt{\pi}. \qquad [15.30]$$

The differential operator was moved to the left by partially integrating n times. Finally, with c_n computed from [15.29] and [15.30], the solution [15.19] reads

$$h_n(x) = \frac{1}{\sqrt{\sqrt{\pi} \cdot 2^n n!}} \cdot \exp\left[-\frac{x^2}{2}\right] H_n(x); \qquad [15.31]$$

a phase factor for this solution can still be chosen arbitrarily. From [15.27] and [15.31], we immediately obtain

$$x h_n(x) = h_{n+1}(x) \sqrt{\frac{n+1}{2}} + h_{n-1}(x) \sqrt{\frac{n}{2}}; \qquad [15.32]$$

and, from [15.25], [15.31], and [15.32], we obtain

$$h_n'(x) = \sqrt{2n}\, h_{n-1}(x) - x h_n(x)$$

$$= -h_{n+1}(x) \sqrt{\frac{n+1}{2}} + h_{n-1}(x) \sqrt{\frac{n}{2}}. \qquad [15.33]$$

The completeness of the Hermite polynomials

Now we want to show that the set of functions $h_n(x)$ is complete; that is, we want to show that Eq. [15.10],

$$y'' - 2xy' + (\lambda - 1)y = 0 , \qquad [15.34]$$

has no regular solutions other than the $h_n(x)$ (that is, no regular solutions except for $\lambda = 2n+1$ and n integral). To this end we look for the general solution of [15.34] and show that it is regular only for $\lambda = 2n+1$. We take a power series,

$$y = \sum_p a_p x^p , \qquad [15.35]$$

as a trial solution. Since there is no singularity at $x = 0$, there are no negative exponents in [15.35].

At this point we want to formulate a general theorem which is useful for the wave equation $\underline{H}\psi = E\psi$: If H is an even function of the p's and x's, as is normally the case when the interaction potential is even $\big(V(x) = V(-x)\big)$, then every solution of the wave equation is the sum of an even solution and an odd solution (as is easily seen with the aid of the wave equation).

In particular, our Hamiltonian function [15.6] is even, and, since $u = e^{-x^2/2}y$, our theorem also applies to the solution of [15.34]; that is, [15.35] can be written as the sum of two power series solutions with

$$p = 0, 2, 4, \ldots \qquad [15.36]$$

and

$$p = 1, 3, 5, \ldots . \qquad [15.37]$$

By substituting [15.35] into [15.34] and comparing the coefficients of x^{p-2}, we obtain the two-term recursion formula

$$a_p \cdot p(p-1) + a_{p-2}\{-2(p-2) + \lambda - 1\} = 0 . \quad [15.38]$$

Let p_0 be the smallest exponent; then, according to [15.38], we have

$$p_0(p_0 - 1) = 0: \qquad \underbrace{p_0 = 0}_{\text{even solution}} \quad \text{or} \quad \underbrace{p_0 = 1}_{\text{odd solution}}.$$

This shows that the initial values in [15.36] and [15.37] are correct. Now we want to investigate if and when the power series [15.35] terminates. The value p_{max} is to be the largest exponent occurring in the power series:

$$a_p = 0 \quad \text{for} \quad p > p_{max} = n; \quad a_n \neq 0. \qquad [15.39]$$

Then, with $p = n+2$ in [15.38],

$$0 = -2n + \lambda - 1 \qquad [15.40]$$

follows; that is, the power series only terminates for

$$\lambda = 2n + 1. \qquad [15.41]$$

For a given odd value of λ, either the even solution terminates or the odd solution terminates, depending on whether n is even or odd. Thus, there is exactly one polynomial; for any other λ neither solution terminates. Now, we want to clarify the convergence properties of the non-terminating series. From [15.38],

$$a_p = a_{p-2} \cdot \frac{2p - 3 - \lambda}{p(p-1)}, \qquad [15.42]$$

follows

$$a_p = a_0 \frac{(1 - \lambda)(5 - \lambda)(9 - \lambda) \cdot \ldots \cdot (2p - 3 - \lambda)}{p!} \qquad [15.43]$$

for p even, and

$$a_p = a_1 \frac{(3 - \lambda)(7 - \lambda) \cdot \ldots \cdot (2p - 3 - \lambda)}{p!} \qquad [15.44]$$

for p odd. For sufficiently large p_M, we certainly have

$$p_M \geq \frac{\lambda + 3}{2}; \qquad [15.45]$$

thus, the coefficients [15.43] and [15.44] no longer change

sign for $p \geqslant p_M$ and we may choose $a_p \geqslant 0$. Further,

$$\lim_{p \to \infty} p \cdot \frac{a_p}{a_{p-2}} = 2 \qquad [15.46]$$

follows from [15.42]. As a simplification we introduce

$$p = 2q \quad \text{or} \quad p = 2q + 1 \quad (q = 0, 1, \ldots) \; ; \quad [15.47]$$

then the series [15.35] becomes

$$y = \sum_q a_{2q} x^{2q} \quad \text{or} \quad y = x \sum_q a_{2q+1} x^{2q} . \qquad [15.48]$$

We shall not be concerned further with the factor x and with the difference between a_{2q+1} and a_{2q} in the right-hand formula. From [15.46],

$$\lim_{q \to \infty} q \cdot \frac{a_{2q}}{a_{2q-2}} = 1 ,$$

we have

$$q \cdot \frac{a_{2q}}{a_{2q-2}} \geqslant \delta \quad \text{or} \quad a_{2q} \geqslant \frac{a_{2q-2} \cdot \delta}{q} \qquad [15.49]$$

for given $\delta < 1$ and sufficiently large q. By repeated application of these expressions, we obtain

$$a_{2q} \geqslant \frac{c \cdot \delta^q}{q!} \; ; \qquad [15.50]$$

using this in [15.48] we can write

$$y = \sum_q a_{2q} x^{2q} \geqslant c \cdot \sum_q \frac{(\delta \cdot x^2)^q}{q!} = c \cdot \exp[+\delta x^2] \qquad [15.51]$$

and

$$u = \exp\left[-\frac{x^2}{2}\right] y \geqslant c \cdot \exp\left|\left(\delta - \frac{1}{2}\right) x^2\right|. \qquad [15.52]$$

Since we can choose δ arbitrarily close to 1 and thus certainly greater than $\frac{1}{2}$, the normalization integral of u does not exist; that is, the nonterminating power series do not provide normalizable solutions of [15.7]. The $h_n(x)$ are the only regular solutions of Eq. [15.7], and this concludes the proof of their completeness.

16. MATRIX CALCULUS ILLUSTRATED WITH THE LINEAR HARMONIC OSCILLATOR

If we have a complete, orthonormal set of functions $\{u_n\}$,

$$\int u_k^* u_l \, dx = \delta_{kl}, \qquad [16.1]$$

then we define the matrix element of the operator $\underline{F}$ with respect to the u_n as

$$(k \,|\, \underline{F} \,|\, n) = \int u_k^* \underline{F} u_n \, dx. \qquad [16.2]$$

The quantity $(k|\underline{F}|n)$ is the element in the kth row and the nth column of the matrix of $\underline{F}$; this element is also designated as $F_{k,n}$. In this way we have associated the operator $\underline{F}$ with a matrix which we also designate by $\underline{F}$. This notation will be justified in the following by the equivalence of matrices and operators.

By means of [16.2] the condition [12.5] for Hermiticity of an operator $\underline{F}$ can be written as

$$(k \,|\, \underline{F} \,|\, n) = (n \,|\, \underline{F} \,|\, k)^*. \qquad [16.3]$$

By analogy, we call a matrix satisfying this condition Hermitian.

Because of [16.1] and [16.2] we can make the following expansion:

$$\underline{F} u_n(x) = \sum_k u_k(x)(k \,|\, \underline{F} \,|\, n). \qquad [16.4]$$

For example, for the operator $\underline{x}$ we have

$$\underline{x} u_n(x) = \sum_k u_k(x)(k \,|\, \underline{x} \,|\, n), \qquad (k \,|\, \underline{x} \,|\, n) = (n \,|\, \underline{x} \,|\, k)^*; \qquad [16.5]$$

and for the operator $\underline{p}_x$ we have

$$- i \frac{du_n}{dx} = \sum_k u_k(x)(k \,|\, \underline{p}_x \,|\, n), \qquad (k \,|\, \underline{p}_x \,|\, n) = (n \,|\, \underline{p}_x \,|\, k)^*. \qquad [16.6]$$

The multiplication of operators corresponds to the multiplication of the associated matrices; an example is

$$\underline{p}_x\{\underline{x}u_n(x)\} = \sum_k - i\frac{du_k}{dx}(k\,|\,\underline{x}\,|\,n)$$

$$= \sum_l u_l(x)\sum_k (l\,|\,\underline{p}_x\,|\,k)(k\,|\,\underline{x}\,|\,n) = \sum_l u_l(x)(l\,|\,\underline{p}_x\underline{x}\,|\,n). \quad [16.7]$$

This agrees exactly with the usual definition of a matrix product:

$$(l\,|\,\underline{A}\underline{B}\,|\,n) \equiv \sum_k (l\,|\,\underline{A}\,|\,k)(k\,|\,\underline{B}\,|\,n). \quad [16.8]$$

Of course, the relation

$$\underline{x}\big(\underline{p}_x u_n(x)\big) = \sum_l u_l(x)(l\,|\,\underline{x}\underline{p}_x\,|\,n) \quad [16.9]$$

also holds, which means that the commutation relation [15.8] takes the form

$$\underline{p}_x\underline{x} - \underline{x}\underline{p}_x = (-i)1 \quad [16.10]$$

in the matrix representation $\big(1$ is the unit matrix, $(k\,|\,1\,|\,l) = \delta_{kl}\big)$.

By comparing [15.32] with the expansion [16.5], we see that only two of the matrix elements of the operator $\underline{x}$ are nonzero if we expand in terms of the eigenfunctions $h_n(x)$:

$$(n+1\,|\,\underline{x}\,|\,n) = \sqrt{\frac{n+1}{2}}, \quad [16.11]$$

$$(n-1\,|\,\underline{x}\,|\,n) = \sqrt{\frac{n}{2}}. \quad [16.12]$$

Because of the Hermiticity of $\underline{x}$, [16.12] follows from [16.11] and vice versa:

$$(n\,|\,\underline{x}\,|\,n+1) = (n+1\,|\,\underline{x}\,|\,n)^* = \sqrt{\frac{n+1}{2}}. \quad [16.13]$$

The relation [16.12] follows from this by means of the substitution $n \to n-1$. We obtain the matrix element of $\underline{p}_x$

from [15.33]:

$$p_x h_n(x) = (-i)\frac{\mathrm{d}h_n(x)}{\mathrm{d}x} = i\sqrt{\frac{n+1}{2}}\,h_{n+1}(x)$$

$$-i\sqrt{\frac{n}{2}}\,h_{n-1}(x) = \sum_k h_k(x)(k\,|\,p_x\,|\,n)\,; \qquad [16.14]$$

thus,

$$(n+1\,|\,p_x\,|\,n) = i\sqrt{\frac{n+1}{2}}\,, \qquad [16.15]$$

$$(n-1\,|\,p_x\,|\,n) = -i\sqrt{\frac{n}{2}}\,. \qquad [16.16]$$

Written out, the matrices are

$$x = \frac{1}{\sqrt{2}}\begin{Vmatrix} 0 & \sqrt{1} & 0 & 0 & \vdots \\ \sqrt{1} & 0 & \sqrt{2} & 0 & \cdot \\ 0 & \sqrt{2} & 0 & \sqrt{3} & \cdot \\ 0 & 0 & \sqrt{3} & 0 & \cdot \\ \cdots\cdots\cdots\cdots\cdots \end{Vmatrix}$$

$$p_x = \frac{1}{i\sqrt{2}}\begin{Vmatrix} 0 & \sqrt{1} & 0 & 0 & \vdots \\ -\sqrt{1} & 0 & \sqrt{2} & 0 & \cdot \\ 0 & -\sqrt{2} & 0 & \sqrt{3} & \vdots \\ 0 & 0 & -\sqrt{3} & 0 & \cdot \\ \cdots\cdots\cdots\cdots\cdots \end{Vmatrix}$$

[16.17]

Note that the indices begin with zero. With

$$x^2 = \frac{1}{2}\begin{Vmatrix} 1 & 0 & \sqrt{2} & 0 & 0 & \vdots \\ 0 & 3 & 0 & \sqrt{6} & 0 & \cdot \\ \sqrt{2} & 0 & 5 & 0 & \sqrt{12} & \vdots \\ 0 & \sqrt{6} & 0 & 7 & 0 & \cdot \\ 0 & 0 & \sqrt{12} & 0 & 9 & \vdots \\ \cdots\cdots\cdots\cdots\cdots\cdots\cdots \end{Vmatrix}$$

$$p_x^2 = \frac{1}{2}\begin{Vmatrix} 1 & 0 & -\sqrt{2} & 0 & 0 & \vdots \\ 0 & 3 & 0 & -\sqrt{6} & 0 & \cdot \\ -\sqrt{2} & 0 & 5 & 0 & -\sqrt{12} & \vdots \\ 0 & -\sqrt{6} & 0 & 7 & 0 & \vdots \\ 0 & 0 & -\sqrt{12} & 0 & 9 & \cdot \\ \cdots\cdots\cdots\cdots\cdots\cdots\cdots \end{Vmatrix}$$

, [16.18]

we obtain

$$(\hbar\omega_0)^{-1}\underline{H} = \tfrac{1}{2}(\underline{x}^2 + \underline{p}_x^2) = \begin{Vmatrix} \tfrac{1}{2} & 0 & 0 & 0 & \cdot \\ 0 & \tfrac{3}{2} & 0 & 0 & \cdot \\ 0 & 0 & \tfrac{5}{2} & 0 & \cdot \\ 0 & 0 & 0 & \tfrac{7}{2} & \cdot \\ \cdot & \cdot & \cdot & \cdot & \cdot \end{Vmatrix} \qquad [16.19]$$

for the Hamiltonian matrix. As a result of expansion with respect to the eigenfunctions, the Hamiltonian matrix is diagonal, and the diagonal elements are the energy eigenvalues. This is true not only in the above example, but, rather, it holds more generally, as we shall see later.

For some applications it is also useful to introduce the following matrices:

$$\underline{A} = \frac{1}{\sqrt{2}}(\underline{x} + i\underline{p}_x) = \frac{1}{\sqrt{2}}\left(x + \frac{\mathrm{d}}{\mathrm{d}x}\right) = \begin{Vmatrix} 0 & \sqrt{1} & 0 & 0 & \cdot \\ 0 & 0 & \sqrt{2} & 0 & \cdot \\ 0 & 0 & 0 & \sqrt{3} & \cdot \\ \cdot & \cdot & \cdot & \cdot & \cdot \end{Vmatrix},$$

$$\underline{A}^* = \frac{1}{\sqrt{2}}(\underline{x} - i\underline{p}_x) = \frac{1}{\sqrt{2}}\left(x - \frac{\mathrm{d}}{\mathrm{d}x}\right) = \begin{Vmatrix} 0 & 0 & 0 & 0 & \cdot \\ \sqrt{1} & 0 & 0 & 0 & \cdot \\ 0 & \sqrt{2} & 0 & 0 & \cdot \\ 0 & 0 & \sqrt{3} & 0 & \cdot \\ \cdot & \cdot & \cdot & \cdot & \cdot \end{Vmatrix}.$$

Then the relation

$$(k\,|\,\underline{A}^*\,|\,n) = (n\,|\,\underline{A}\,|\,k)^*$$

holds, and the relation

$$[\underline{A},\,\underline{A}^*] \equiv \underline{A}\,\underline{A}^* - \underline{A}^*\underline{A} = 1$$

follows from the commutation relations [16.10].

a. Equivalence of the Schrödinger equation to a system of equations in Hilbert space

According to [9.2], we can expand an arbitrary quadratically integrable function $\psi(x)$ in terms of a complete orthonormal set of functions $u_n(x)$:

$$\psi(x) = \sum u_n(x)\psi_n , \qquad \psi_n = \int \psi(x) u_n^*(x)\,\mathrm{d}x . \qquad [16.20]$$

In this way, a vector with components ψ_n in an infinite-dimensional Hilbert space is associated with each function $\psi(x)$. In this case the completeness relation is

$$\int |\psi(x)|^2\mathrm{d}x = \sum_k |\psi_k|^2 , \qquad [16.21]$$

from which follows the existence of $\sum\limits_k |\psi_k|^2$. Therefore, instead of calculating with wave functions, one can just as well calculate with vectors in Hilbert space, applying the rules of matrix multiplication.

If we substitute the expansion [16.20] into the Schrödinger equation, multiply by $u_k^*(x)$, and integrate, we then obtain

$$\int \mathrm{d}x u_k^*(x) \sum_n \underline{H} u_n(x) \cdot \psi_n = \int \mathrm{d}x u_k^*(x) \sum_n E_n u_n(x) \cdot \psi_n . \qquad [16.22]$$

Combined with [16.2] and [16.4], this yields

$$\sum_n \{(k\,|\,\underline{H}\,|\,n) - E_n(k\,|\,1\,|\,n)\}\psi_n = 0 . \qquad [16.23]$$

This linear, homogeneous system of infinitely many equations ($k = 0, 1, 2, \ldots$) in the unknowns ψ_n is exactly equivalent to the Schrödinger equation and gives the same results.

b. Example: Linear oscillator with an additional potential

With the following considerations we want to prepare the perturbation theory. If an additional potential $V(x)$ is added to the Hamiltonian function H_0 of the linear harmonic oscillator (we put $b\omega_0 = 1$ for the rest of this section),

$$\underline{H} = \tfrac{1}{2}(\underline{x}^2 + \underline{p}_x^2) + \underline{V}(x) = \underline{H}_0 + \underline{V}(x) , \qquad [16.24]$$

then the corresponding Schrödinger equation,

$$\underline{H}\psi = E\psi\,,$$
or
$$(\underline{H} - E\cdot 1)\psi = 0\,,$$

in matrix form, is usually no longer exactly solvable. If the additional potential is small (a "perturbation"), then an approximate solution can be found which is especially easy to calculate in the matrix representation.

Since the $h_n(x)$ are complete and orthonormal, we can make the expansions

$$\psi(x) = \sum_n h_n(x)\psi_n\,, \qquad \psi_n = \int \psi(x)h_n^*(x)\,\mathrm{d}x\,, \qquad [16.25]$$

$$\left.\begin{aligned} V(x)h_n(x) &= \sum_k h_k(x)\,(k\,|\,\underline{V}\,|\,n) \\ (k\,|\,\underline{V}\,|\,n) &= \int h_k^*(x)\,V(x)h_n(x)\,\mathrm{d}x \end{aligned}\right\}\,, \qquad [16.26]$$

and there is the relation

$$\int h_k^*(x)h_n(x)\,\mathrm{d}x = (k\,|\,1\,|\,n)\,. \qquad [16.27]$$

As we did with [16.22], we substitute this expansion into the Schrödinger equation, multiply by $h_k^*(x)$, and integrate:

$$\int \mathrm{d}x\, h_k^*(x) \sum_n \{(\underline{H_0} - E)h_n(x) + \underline{V}(x)h_n(x)\}\psi_n = 0\,.$$

We then obtain

$$\sum_n \{(n + \tfrac{1}{2} - E)(k\,|\,1\,|\,n) + (k\,|\,\underline{V}\,|\,n)\}\psi_n = 0\,. \qquad [16.28]$$

This is again a system of infinitely many equations in the unknowns ψ_n which is equivalent to the Schrödinger equation. We shall solve the system later when we consider perturbation theory.

c. Determination of the eigenvalues of the linear harmonic oscillator with the matrix method

We have just derived the matrix representation of quantum mechanics from the wave equation, and we have shown that the two types of representation are completely equivalent. Nevertheless, historically, matrix mechanics was developed before wave mechanics,[2] and at first the identity of the two theories was not at all noticed.

As an example of a matrix mechanical calculation without recourse to the wave equation, we treat the linear harmonic oscillator again. Thus, in the Hamiltonian matrix for the linear harmonic oscillator,

$$\underline{H} = \tfrac{1}{2}(\underline{p}_x^2 + \underline{x}^2) \,, \qquad [16.29]$$

where, for example, $\underline{x}^2$ stands for the matrix

$$(n \,|\, \underline{x}^2 \,|\, n') = \sum_k (n \,|\, \underline{x} \,|\, k)(k \,|\, \underline{x} \,|\, n') \,,$$

we must substitute Hermitian matrices for $\underline{x}$ and $\underline{p}_x$ which make $\underline{H}$ diagonal. In addition, $\underline{x}$ and $\underline{p}_x$ must obey the commutation relations [13.13] and [13.14], which, of course, can be derived within the matrix mechanical framework. In our special coordinates [15.4], the commutation relations are

$$\underline{H}\underline{p}_x - \underline{p}_x\underline{H} = i\,\frac{\partial \underline{H}}{\partial \underline{x}} = +i\underline{x} \,, \qquad [16.30]$$

$$\underline{H}\underline{x} - \underline{x}\underline{H} = -i\,\frac{\partial \underline{H}}{\partial \underline{p}_x} = -i\underline{p}_x \,. \qquad [16.31]$$

---

[2] W. HEISENBERG (*Z. Physik* **33**, 879 (1925)) introduced the matrix elements as the quantum mechanical analogues of the Fourier amplitudes of classical mechanics. Just as a classical quantity is determined by its Fourier amplitudes, so the associated quantum mechanical quantity is to be given by the set of associated matrix elements (although with the matrix elements a form corresponding to the Fourier integral is not possible). However, Heisenberg did not yet use the expression "matrix element" in his first article. Born and Jordan (*Z. Physik* **34**, 858 (1925)) were the first to recognize that the law of multiplication given by Heisenberg for the quantum mechanical quantities was identical to the rule for matrix multiplication. By means of the matrix calculus the entire theory could then be still better founded and worked out (M. BORN, W. HEISENBERG and P. JORDAN, *Z. Physik* **35**, 557 (1926)).

If we now assume that $\underline{H} = \underline{E}$ is already diagonal, as a result of the correct choice of $\underline{x}$ and $\underline{p}_x$, then such a commutator becomes simply

$$(n' | \underline{H}\underline{F} - \underline{F}\underline{H} | n'') = (E_{n'} - E_{n''})(n' | \underline{F} | n'')$$

(Pauli: "... just calculate it!"), which, with [16.30] and [16.31], yields

$$(E_{n'} - E_{n''})(n' | \underline{p}_x | n'') = i(n' | \underline{x} | n'') ,$$

$$(E_{n'} - E_{n''})(n' | \underline{x} | n'') = - i(n' | \underline{p}_x | n'').$$

Therefore, $(n' | \underline{p}_x | n'')$ and $(n' | \underline{x} | n'')$ are either both zero or they are both different from zero; in the latter case, as a result of

$$(E_{n'} - E_{n''})^2 (n' | \underline{x} | n'') = (n' | \underline{x} | n'') ,$$

we have

$$(E_{n'} - E_{n''})^2 = 1 , \quad \text{or} \quad E_{n'} - E_{n''} = \pm 1 . \qquad [16.32]$$

Since the eigenvalues of the linear harmonic oscillator are nondegenerate, as can also be easily shown within the framework of matrix mechanics, the selection rule [3]

$$n' - n'' = \pm 1 \qquad\qquad [16.33]$$

follows from [16.32]; simultaneously we have fixed the order in which the eigenvalues are to be numbered:

$$E_n = n + \text{constant} . \qquad\qquad [16.34]$$

With this numbering, only the matrix elements

$$(n | \underline{x} | n+1) , \quad (n | \underline{x} | n-1) , \quad (n | \underline{p}_x | n+1) , \quad (n | \underline{p}_x | n-1)$$

are different from zero; all others equal zero. From the

---

[3] Only those matrix elements whose indices obey the "selection rule" are different from zero. We shall see later that the intensity of light emission is proportional to the square of the matrix element associated with the transition responsible for the emission. Thus, the selection rule specifies between which states transitions are possible.

above relation,

$$(n \,|\, \underline{p}_x \,|\, n \mp 1) = \pm i(n \,|\, \underline{x} \,|\, n \mp 1) \qquad [16.35]$$

now follows.

We have not yet been completely justified in writing Eq. [16.34]. We must first show that the sequence formed by the E_n has no holes; that is, we must show that the next larger energy following each E_n is $E_{n+1} = E_n + 1$. For this purpose we consider the diagonal element of the commutator [16.10],

$$i(n \,|\, \underline{p}_x \underline{x} - \underline{x}\underline{p}_x \,|\, n) = 2\{(n \,|\, \underline{x} \,|\, n+1)(n+1 \,|\, \underline{x} \,|\, n)$$
$$- (n \,|\, \underline{x} \,|\, n-1)(n-1 \,|\, \underline{x} \,|\, n)\} = 1 \ ; \qquad [16.36]$$

in arriving at this expression for the commutator we have used [16.35]. As a result of the Hermiticity of $\underline{x}$, we obtain

$$|(n \,|\, \underline{x} \,|\, n+1)|^2 - |(n \,|\, \underline{x} \,|\, n-1)|^2 = \tfrac{1}{2} \qquad [16.37]$$

from [16.36]. That is, we have

$$|(n \,|\, \underline{x} \,|\, n-1)|^2 = \frac{n}{2} + \text{constant} , \qquad [16.38]$$

since this implies

$$|(n \,|\, \underline{x} \,|\, n+1)|^2 = |(n+1 \,|\, \underline{x} \,|\, n)|^2 = \frac{n+1}{2} + \text{constant}$$

which then leads to [16.37]. Using formula [16.35] and the Hermiticity of $\underline{x}$ and $\underline{p}_x$, we also obtain

$$(n \,|\, \underline{p}_x^2 \,|\, n) = (n \,|\, \underline{x}^2 \,|\, n)$$
$$= |(n \,|\, \underline{x} \,|\, n-1)|^2 + |(n \,|\, \underline{x} \,|\, n+1)|^2, \qquad [16.39]$$

with which we then obtain

$$E_n = |(n \,|\, x \,|\, n-1)|^2 + |(n \,|\, x \,|\, n+1)|^2 = n + \text{constant} \qquad [16.40]$$

from [16.29]. The left side of this equation is again positive. Therefore, there must be a smallest $n = n_0$ such that $E_n \equiv 0$ for $n < n_0$. Since we have only decided about the sequence of the numbering until now, we can take $n_0 = 0$. Then we have $E_{-1} = 0$, which means

$$(-1 \mid \underline{x} \mid 0) = 0 ;$$

together with [16.37], this yields

$$E_0 = \tfrac{1}{2} \text{ and } E_n = n + \tfrac{1}{2} . \qquad [16.41]$$

The E_0 is the famous *zero-point energy* which we now have derived correctly with the matrix method also (see [15.20]).

17. THE HARMONIC OSCILLATOR IN A PLANE. DEGENERACY

The isotropic harmonic oscillator in a plane has the potential

$$V(q_1, q_2) = \frac{m}{2} \omega_0^2 (q_1^2 + q_2^2) , \qquad [17.1]$$

which leads to the Schrödinger equation

$$\frac{d^2 u}{dx_1^2} + \frac{d^2 u}{dx_2^2} + (\lambda - x_1^2 - x_2^2) u = 0 , \qquad [17.2]$$

with

$$\lambda = 2E/\hbar\omega_0 \qquad \text{and} \qquad x_i = \sqrt{m\omega_0/\hbar} \cdot q_i$$

$$\big(\text{see } [15.4] \text{ and } [15.5]\big).$$

Equation [17.2] is the sum of two equations in x_1 and x_2. Therefore, the solution reduces to the product

$$u = h_{n_1}(x_1) \cdot h_{n_2}(x_2) , \qquad [17.3]$$

and the eigenvalues are

$$\begin{aligned}
\lambda = \lambda_1 + \lambda_2 &= 2n_1 + 1 + 2n_2 + 1 \\
&= 2(n_1 + n_2 + 1) = 2(n + 1) .
\end{aligned} \qquad [17.4]$$

(When the solution is a product of this form, we say that the equation is *separable*.) Here we have the situation in which several different states are associated with the same eigenvalue; this is known as *degeneracy*. If n states belong to the same eigenvalue, we speak of n-fold degeneracy. In our case the degeneracy is $(n+1)$-fold:

$$n_1: \quad 0, \quad 1, \quad 2, ..., n ,$$
$$n_2: \quad n, n-1, n-2, ..., 0 .$$

We can remove this degeneracy if we appropriately modify the potential—for example, by going to the anisotropic harmonic oscillator:

$$V(q_1, q_2) = \frac{m}{2}(\omega_1^2 q_1^2 + \omega_2^2 q_2^2) . \quad [17.5]$$

The equation is still separable and the solution is

$$u = h_{n_1}\left(\sqrt{\frac{m\omega_1}{h}}\, q_1\right) \cdot h_{n_2}\left(\sqrt{\frac{m\omega_2}{h}}\, q_2\right), \quad [17.6]$$

with

$$E = h\{\omega_1(n_1 + \tfrac{1}{2}) + \omega_2(n_2 + \tfrac{1}{2})\} . \quad [17.7]$$

If ω_1/ω_2 is irrational (ω_1 and ω_2 incommensurable), then the degeneracy is removed.

By going to the limit $\omega_1/\omega_2 \to 1$, we again obtain the isotropic harmonic oscillator. Quite generally, a degenerate system can be thought of as obtained, usually as a limiting case, from various separable nondegenerate systems; that is, the degenerate system will be separable in various coordinate systems.

a. The solution of the plane harmonic oscillator in polar coordinates

For example, the isotropic harmonic oscillator is also separable in polar coordinates:

$$x_1 = r\cos\varphi, \quad x_2 = r\sin\varphi, \quad dx_1 dx_2 = r\, dr\, d\varphi, \quad [17.8]$$

$$\left(\frac{\partial^2}{\partial x_1^2} + \frac{\partial^2}{\partial x_2^2}\right) u = \frac{1}{r}\frac{\partial}{\partial r}\left(r\frac{\partial u}{\partial r}\right) + \frac{1}{r^2}\frac{\partial^2 u}{\partial \varphi^2}$$

$$= \frac{\partial^2 u}{\partial r^2} + \frac{1}{r}\frac{\partial u}{\partial r} + \frac{1}{r^2}\frac{\partial^2 u}{\partial \varphi^2} . \quad [17.9]$$

The Schrödinger equation is then

$$\frac{\partial^2 u}{\partial r^2} + \frac{1}{r}\frac{\partial u}{\partial r} + \frac{1}{r^2}\frac{\partial^2 u}{\partial \varphi^2} + (\lambda - r^2)u = 0. \qquad [17.10]$$

The solution can be separated as a product of the form

$$u = v_m(r)\,e^{im\varphi}, \qquad [17.11]$$

where m is a positive or negative integer; then the differential equation for $v_m(r)$ is

$$\frac{d^2 v_m}{dr^2} + \frac{1}{r}\frac{dv_m}{dr} - \frac{m^2}{r^2}v_m + (\lambda - r^2)v_m = 0. \qquad [17.12]$$

Since the energy values of a system are not changed by a coordinate transformation, the degeneracy must be the same as in [17.4]; on the other hand, the eigenfunctions will be different. Indeed, even in one coordinate system they can be made quite different by taking linear combinations. We can also remove the degeneracy by going to the isotropic anharmonic oscillator:

$$(h\omega_0)^{-1}V(r) = \tfrac{1}{2}r^2 + \varepsilon\,\overline{V}(r). \qquad [17.13]$$

The ε is to be a small number which is not proportional to r^2. The separability [17.11] is not altered and we obtain

$$\frac{d^2 v_m}{dr^2} + \frac{1}{r}\frac{dv_m}{dr} - \frac{m^2}{r^2}v_m + \left(\lambda - r^2 - 2\varepsilon\,\overline{V}(r)\right)v_m = 0 \qquad [17.14]$$

in analogy to [17.12]. The eigenvalues of this equation are still two-fold degenerate for the values $\pm m$; in order to remove this degeneracy we would have to introduce a magnetic field.

Now, we want to solve Eq. [17.12]. Since the physically allowed range of r is $0 \leqslant r \leqslant \infty$, we can simplify [17.12] by the coordinate transformation

$$r^2 = x, \qquad r = \sqrt{x}, \qquad r\frac{\partial}{\partial r} = 2x\frac{\partial}{\partial x}; \qquad [17.15]$$

we then obtain

$$\frac{1}{r}\frac{\partial}{\partial r}\left(r\frac{\partial u}{\partial r}\right) = 2\frac{\partial}{\partial x}\left(2x\frac{\partial u}{\partial x}\right) = 4\left(x\frac{\partial^2 u}{\partial x^2} + \frac{\partial u}{\partial x}\right),$$

$$x\frac{d^2 v_m}{dx^2} + \frac{dv_m}{dx} - \frac{m^2}{4x}v_m + \frac{\lambda - x}{4}v_m = 0 . \qquad [17.16]$$

With $dy/dx = y'$ and the substitutions

$$v_m = x^{|m|/2}e^{-x/2}y , \qquad [17.17]$$

$$v'_m = e^{-x/2}x^{|m|/2}\left\{y' + y\left(\frac{m}{2x} - \frac{1}{2}\right)\right\},$$

$$v''_m = e^{-x/2}x^{|m|/2}\left\{y'' + 2\left(\frac{m}{2x} - \frac{1}{2}\right)y' - \frac{m}{2x^2}y + \left(\frac{m}{2x} - \frac{1}{2}\right)^2 y\right\},$$

we can write Eq. [17.16] as

$$y'' + \left(\frac{m+1}{x} - 1\right)y' + \frac{1}{2x}\left(\frac{\lambda}{2} - m - 1\right)y = 0 \qquad [17.18]$$

or

$$xy'' + (m+1-x)y' + ky = 0 , \qquad [17.19]$$

where

$$k = \frac{1}{2}\left(\frac{\lambda}{2} - m - 1\right).$$

As a result of the relation [17.4], we also have

$$n = 2k + m .$$

We treat the case of m and k integral and nonnegative. The treatment of the general differential equation [17.19] yields physically useful solutions only for $k = 0, 1, 2, \ldots$.

In the following we want to discuss some functions which solve [17.19].

b. *The Laguerre polynomials*

The Laguerre polynomials are defined by

$$L_k(x) = e^x \frac{\mathrm{d}^k}{\mathrm{d}x^k} \left(x^k e^{-x} \right) = \sum_{n=0}^{k} (-1)^n \binom{k}{n}$$

$$\times \, k(k-1) \cdot \ldots \cdot (n+1) \cdot x^n, \qquad [17.20]$$

and they satisfy the differential equation [17.19] with $m = 0$,

$$x L_k'' + (1-x) L_k' + k L_k = 0 . \qquad [17.21]$$

This can be shown, for example, with the generating function for the Laguerre polynomials,

$$f(x, z) = \sum_{k=0}^{\infty} \frac{L_k(x)}{k!} z^k = \frac{\exp \left[- xz/(1-z) \right]}{1-z} ,$$

which we derive generally in [17.35]. Differentiation with respect to x and z and comparison of coefficients yields

$$L_k' - k L_{k-1}' = - k L_{k-1} , \qquad [17.22]$$

$$L_{k+1} = (2k+1-x) L_k - k^2 L_{k-1} . \qquad [17.23]$$

The differential equation [17.21] follows immediately from this (substitute L_{k-1} from [17.23] into [17.22], and eliminate L_{k+1} with the relation $L_{k+1}' - (k+1) L_k' = -(k+1) L_k$).
With the definition

$$L_k^m \equiv \frac{\mathrm{d}^m}{\mathrm{d}x^m} L_k , \qquad [17.24]$$

we obtain

$$x (L_k^m)'' + (m+1-x)(L_k^m)' + (k-m) L_k^m = 0 \qquad [17.25]$$

from [17.21] by differentiating m times. If we replace k in this equation by $k+m$, then we obtain exactly the differential equation [17.19], which proves that

$$L_{k+m}^m \equiv \frac{\mathrm{d}^m}{\mathrm{d}x^m} L_{k+m} \qquad [17.26]$$

satisfies Eq. [17.19] for $m \geqslant 0$.

Therefore, the solution of Eq. [17.16] (and the corresponding Eq. [17.12] with which we began) is

$$v_{k,m} = \text{constant} \times x^{m/2}\, e^{-x/2}\, L^m_{k+m}(x). \qquad [17.27]$$

We shall normalize this solution in [17.38] and generalize it to the hypergeometric functions in [17.37].

The L^m_{k+m} are a set of orthogonal functions; that is, they satisfy

$$\int L^m_{k+m} L^m_{k'+m}\,\mathrm{d}x = 0 \qquad \text{for} \qquad k \neq k', \qquad [17.28]$$

of which anyone can convince himself by performing the calculation.

By applying the formula

$$f^{(k)}(x) = \frac{k!}{2\pi i} \oint_C \frac{f(t)}{(t-x)^{k+1}}\, \mathrm{d}t, \qquad [17.29]$$

where $k > 0$ and C is a circle about x, to $f(x) = x^k e^{-x}$, and using [17.20], we obtain

$$L_k(x) = e^x k! \frac{1}{2\pi i} \oint_C \frac{e^{-t} t^k}{(t-x)^{k+1}}\, \mathrm{d}t. \qquad [17.30]$$

By substituting $-t+x$ for t, this can be written as

$$L_k(x) = \frac{k!}{2\pi i} \oint_C e^t (t-x)^k t^{-(k+1)}\mathrm{d}t. \qquad [17.31]$$

Replacing k by $k+m$ and differentiating m times with respect to x, we also obtain an analogous expression for L^m_{k+m}:

$$L^m_{k+m} = \frac{\mathrm{d}^m}{\mathrm{d}x^m} L_{k+m} = (-1)^m \frac{[(k+m)!]^2}{k!}$$

$$\times \frac{1}{2\pi i} \oint_C e^t (t-x)^k t^{-(k+m+1)}\mathrm{d}t. \qquad [17.32]$$

Using this practical integral representation, we want to derive the generating function for the Laguerre polynomials:

$$f(x, z) = \sum_{k=0}^{\infty} \frac{L_k(x)}{k!} z^k = \frac{1}{2\pi i} \oint_C e^t \sum_{k=0}^{\infty} \left(\frac{t-x}{t} z \right)^k \frac{dt}{t} . \qquad [17.33]$$

We can assume $|z| < 1$; in this way the convergence of the series is guaranteed for sufficiently large t:

$$\left| \frac{t-x}{t} \right| \cdot |z| < 1 .$$

The calculation yields

$$f(x, z) = \frac{1}{2\pi i} \oint e^t \frac{1}{\left(1 - \frac{t-x}{t} z \right)} \frac{dt}{t} = \frac{1}{2\pi i} \oint e^t \frac{dt}{t - (t-x)z}$$

$$= \frac{1}{1-z} \cdot \frac{1}{2\pi i} \oint e^t \frac{dt}{t + \dfrac{xz}{1-z}} . \qquad [17.34]$$

The integrand has a simple pole at the point $t = -(xz)/(1-z)$. Therefore, the integral equals $2\pi i$ times the residue, or

$$2\pi i \exp\left[-\frac{xz}{1-z} \right] ,$$

and the final result is

$$f(x, z) = \sum_{k=0}^{\infty} \frac{L_k(x)}{k!} z^k = \frac{\exp[-xz/(1-z)]}{1-z} . \qquad [17.35]$$

The generating function for L_{k+m}^m follows from this simply by differentiating m times with respect to x:

$$\sum_{k=m}^{\infty} \frac{L_k^m(x)}{k!} z^k = (-1)^m z^m \frac{\exp[-xz/(1-z)]}{(1-z)^{m+1}} .$$

Replacing k by $k+m$ then results in

$$\sum_{k=0}^{\infty} \frac{L_{k+m}^m(x)}{(k+m)!} z^k = (-1)^m \frac{\exp[-xz/(1-z)]}{(1-z)^{m+1}} . \qquad [17.36]$$

c. Normalization of the solution [17.27]

Now, we want to normalize the solution of Eq. [17.16] which is given by [17.27],

$$v_{k,m}(x) = \text{constant} \times x^{m/2} \, e^{-x/2} \, L_{k+m}^m(x) \,, \qquad [17.37]$$

and prove its orthogonality at the same time. We calculate the normalization integral,

$$\int_0^\infty v_{k,m} v_{k',m} \, \mathrm{d}x = \int_0^\infty x^m \, e^{-x} L_{k+m}^m(x) L_{k'+m}^m(x) \, \mathrm{d}x = N_{km} \delta_{kk'} \,, \qquad [17.38]$$

by the elegant method of E. Schrödinger; this method can be applied completely generally (as can the derivation just given of the generating function). To this end we use [17.36] in the form

$$\sum_{k'=0}^\infty \frac{L_{k'+m}^m(x)}{(k'+m)!} t^{k'} = (-1)^m \frac{\exp[-xt/(1-t)]}{(1-t)^{m+1}} \,, \qquad [17.39]$$

and we integrate the product of [17.36], [17.39], and $x^m e^{-x}$ with respect to x:

$$\sum_{k=0}^\infty \sum_{k'=0}^\infty \frac{z^k t^{k'}}{(k+m)!\,(k'+m)!} \int_0^\infty x^m \exp[-x] L_{k+m}^m(x) L_{k'+m}^m(x) \, \mathrm{d}x$$

$$= \frac{1}{(1-z)^{m+1}} \frac{1}{(1-t)^{m+1}} \int_0^\infty x^m \exp\{-x(1-zt)/[(1-t)(1-z)]\} \, \mathrm{d}x$$

$$= \frac{1}{(1-zt)^{m+1}} \int_0^\infty y^m \exp[-y] \, \mathrm{d}y = \frac{m!}{(1-zt)^{m+1}}$$

$$= m! \sum_{k=0}^\infty \binom{-m-1}{k} (-1)^k (zt)^k = \sum_{k=0}^\infty \frac{(k+m)!}{k!} (zt)^k \,. \qquad [17.40]$$

For this calculation we used the following relations:

$$y = x \frac{1 - zt}{(1 - z)(1 - t)}, \qquad \int_0^\infty y^m e^{-y} \, dy = m!,$$

$$(-1)^k \binom{-m-1}{k}$$

$$= \frac{(m+1)(m+2) \cdot \ldots \cdot (m+k)}{k!} = \frac{(m+k)!}{k!\,m!}. \qquad [17.41]$$

By comparing coefficients in [17.40], we find that formula [17.38] is satisfied with

$$N_{k,m} = \frac{[(k+m)!]^3}{k!}. \qquad [17.42]$$

Indeed, the coefficients of all terms of the form $z^k t^{k'}$ with $k \neq k'$ on the right-hand side are zero; the orthogonality follows from that fact.

Remark: We define

$$\binom{\alpha}{n} \equiv \frac{\alpha(\alpha-1) \cdot \ldots \cdot (\alpha - n + 1)}{n!} \qquad [17.43]$$

also for nonintegral values of α. Formula [17.41] is valid in this sense:

$$\binom{-\alpha}{n}(-1)^n = \frac{\alpha(\alpha+1) \cdot \ldots \cdot (\alpha+n-1)}{n!} = \frac{\Gamma(\alpha+n)}{\Gamma(\alpha) \cdot n!}. \qquad [17.44]$$

d. Some properties of the Γ function

Here we gather together some properties of the Γ function which we shall need later. The Γ function satisfies the well-known functional equations

$$\Gamma(z+1) = z \cdot \Gamma(z), \qquad [17.45]$$

$$\Gamma(z) \cdot \Gamma(1-z) = \frac{\pi}{\sin \pi z}. \qquad [17.46]$$

If $z = n$ ($n = 0, 1, 2, ...$), then the first equation with $\Gamma(1) = 1$ reduces to

$$\Gamma(n+1) = n! \,. \qquad [17.47]$$

$\Gamma(z)$ has simple poles at $z = 0, -1, -2, ...$, and it is regular everywhere else; $1/\Gamma(z)$ is an analytic transcendental function. The Γ function has the Euler integral representation

$$\Gamma(z+1) = \int_0^\infty e^{-t} t^z \mathrm{d}t \quad \text{for} \quad \mathrm{Re}\,(z) > -1 \,. \qquad [17.48]$$

In addition, there is the simple, useful, and beautiful Hankel relation,

$$\frac{1}{\Gamma(z)} = \frac{1}{2\pi i} \int_C e^t \, t^{-z} \mathrm{d}t \quad \text{for all } z, \qquad [17.49]$$

where the path of integration C, as shown in Fig. 17.1, goes around the origin and converges toward the real axis

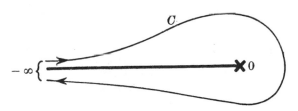

Figure 17.1

at $-\infty$. Either one of the integral representations can be transformed into the other by means of [17.46]. Because of [17.45], the relation

$$\frac{\Gamma(\gamma)}{\Gamma(\gamma+n)} = \frac{1}{\gamma(\gamma+1)\cdot...\cdot(\gamma+n-1)} \qquad [17.50]$$

also holds.

e. The confluent hypergeometric function

The general differential equation (see [17.19], $m+1 \to \gamma$, $k \to -\alpha$)

$$xy'' + (\gamma - x)y' - \alpha y = 0 \,, \qquad [17.51]$$

where α, γ, and x can be arbitrary real or complex quantities, leads to the *confluent hypergeometric function* $F(\alpha, \gamma, x)$, which is a limiting case of the general hypergeometric function $F(\alpha, \beta, \gamma, x)$: [4]

$$F(\alpha, \gamma, x) = \lim_{\beta \to \infty} F\left(\alpha, \beta, \gamma, \frac{x}{\beta}\right).$$

First, we want to derive an integral representation of the hypergeometric function and then investigate its asymptotic behavior. To solve [17.51] we use the series

$$y = \sum_{n=0}^{\infty} a_n x^n \qquad [17.52]$$

as a trial solution. By comparing the coefficients of x^n we obtain the recursion formula

$$(n+1)na_{n+1} + \gamma(n+1)a_{n+1} - na_n - \alpha a_n = 0 \, ,$$

$$a_{n+1} = \frac{\alpha + n}{\gamma + n} \frac{1}{n+1} \cdot a_n \, . \qquad [17.53]$$

In order that the solution not vanish identically, we must assume $a_0 \neq 0$. From [17.53] it is also evident that no solution of the form [17.52] exists for $\gamma \leqslant 0$ and integral. If we set $a_0 = 1$, then the solution is

$$F(\alpha, \gamma, x) = 1 + \frac{\alpha}{1!\gamma} x + \frac{\alpha(\alpha+1)}{2!\gamma(\gamma+1)} x^2 + \dots$$
$$+ \frac{\alpha(\alpha+1) \dots (\alpha+n-1)}{n!\gamma(\gamma+1) \dots (\gamma+n-1)} x^n + \dots . \qquad [17.54]$$

---

[4] The general hypergeometric function is defined by the hypergeometric series

$$F(a, b, c, z) = 1 + \frac{ab}{1!c} z + \frac{a(a+1)\,b(b+1)}{2!\,c(c+1)} z^2 + \dots ,$$

in which all quantities can be complex. The series is absolutely convergent for $|z| < 1$ and divergent for $|z| > 1$; it is also absolutely convergent for $|z| = 1$ if $\mathrm{Re}\,(a + b - c) < 0$. The hypergeometric function satisfies the equation

$$z\,(1-z)\frac{d^2u}{dz^2} + \{c - (a + b + 1)z\}\frac{du}{dz} - abu = 0 \, .$$

For $\alpha = 0, -1, -2, \ldots, -k, \ldots$, the series terminates, which means that $F(-k, \gamma, x)$ is a polynomial; as we shall immediately see, for $\gamma = m+1$ these are essentially the Laguerre polynomials.

Using [17.44] and [17.49] we can transform this solution into an integral:

$$F(\alpha, \gamma, x) = \Gamma(\gamma) \sum_{n=0}^{\infty} \binom{-\alpha}{n} \frac{(-x)^n}{\Gamma(\gamma + n)}$$

$$= \Gamma(\gamma) \frac{1}{2\pi i} \int_C e^t \sum_{n=0}^{\infty} \binom{-\alpha}{n} t^{-\gamma} \left(\frac{-x}{t}\right)^n dt.$$

The requirement

$$\left|\frac{x}{t}\right| < 1, \qquad\qquad [17.55]$$

which must be made in order that the series converge, means that the path of integration C (see formula [17.49]) must surround the point x as well as the origin (Fig. 17.2).

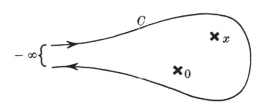

Figure 17.2

Using the binomial theorem,

$$\sum_{n=0}^{\infty} \binom{-\alpha}{n} \left(\frac{-x}{t}\right)^n = \left(1 - \frac{x}{t}\right)^{-\alpha},$$

we obtain

$$F(\alpha, \gamma, x) = \frac{\Gamma(\gamma)}{2\pi i} \int_C e^t \, t^{\alpha - \gamma} (t - x)^{-\alpha} dt. \qquad [17.56]$$

Substituting

$$\gamma = m+1, \qquad \alpha = -k$$

into this formula (in which case [17.51] reduces to Eq. [17.19]), we again obtain the Laguerre polynomials of [17.32]; the only difference is that they were normalized differently:

$$L_{k+1}^m(x) = \frac{[(k+m)!]^2(-1)^m}{k!} \frac{1}{2\pi i} \int_o e^t \, t^{-(k+m+1)}(t-x)^k dt$$

$$= (-1)^m \binom{k+m}{m}(k+m)! \, F(-k, m+1, x) \, . \qquad [17.57]$$

We now want to check that the solution [17.56] which we derived actually satisfies the differential equation [17.51] for the confluent hypergeometric function. To this end we state the identity

$$xF'' + (\gamma - x)\, F' - \alpha F$$
$$\equiv \frac{\Gamma(\gamma)}{2\pi i}\, (-\alpha) \int_o \frac{\mathrm{d}}{\mathrm{d}t}\, [e^t \, t^{\alpha-\gamma+1}(t-x)^{-\alpha-1}]\, \mathrm{d}t \, , \qquad [17.58]$$

which can be derived immediately with the aid of [17.56] and the relations

$$e^t \, t^{\alpha-\gamma}(t-x)^{-\alpha} \left\{ \frac{(-\alpha)(-\alpha-1)}{(t-x)^2}\, x + (\gamma - x)\frac{\alpha}{t-x} - \alpha \right\}$$

$$= e^t \, t^{\alpha-\gamma}(t-x)^{-\alpha-1}(-\alpha) \left\{ (-\alpha-1)\frac{t}{t-x} + (\alpha+1-\gamma) + t \right\}$$

$$= \frac{\mathrm{d}}{\mathrm{d}t} \{ e^t \, t^{\alpha-\gamma+1}(t-x)^{-\alpha-1} \} \, .$$

We know that the path of integration C (see Fig. 17.1) runs from $-\infty$, around the origin, and back to $-\infty$. However, at the end points $-\infty$ the integral vanishes, so that the right-hand side of the identity equals zero. This shows that the equation is satisfied.

If the path of integration C is broken up as shown in Fig. 17.3, then we have

$$\int_C = \int_{C_1} + \int_{C_2},$$
$$\downarrow \qquad \downarrow \qquad \downarrow$$
$$F = F_1 + F_2,$$

with

$$F_1 \atop 2 = \frac{\Gamma(\gamma)}{2\pi i} \int_{C_1 \atop C_2} e^t\, t^{\alpha-\gamma}(t-x)^{-\alpha}\mathrm{d}t\,. \qquad [17.59]$$

According to [17.58], since the end points of the integration are all at $-\infty$, both F_1 and F_2 are solutions of the differential equation [17.51]. Thus, we have split our solution F,

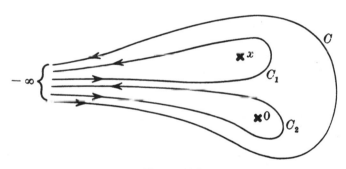

Figure 17.3

given by [17.56], into the two independent solutions F_1 and F_2. If $-\alpha = k = 0, 1, 2, \ldots$, then there is at most one pole at the origin and

$$F_1 = 0\,, \qquad F = F_2\,;$$

for $\gamma = m+1 > -k$ this again yields the Laguerre polynomials. On the other hand, if the origin is a regular point, which means $\alpha - \gamma$ integral and nonnegative, then we have

$$F_2 = 0\,, \qquad F = F_1\,.$$

f. Asymptotic behavior of the confluent hypergeometric function

With the aid of the integrals just introduced we can investigate the behavior of $F(\alpha, \gamma, x)$ for large $|x|$.

First, we calculate F_2. For large $|x|$ we can use the expansion

$$(t - x)^{-\alpha} = (-x)^{-\alpha}\left(1 - \frac{t}{x}\right)^{-\alpha}$$

$$= (-x)^{-\alpha}\left(1 + \alpha\frac{t}{x} + ...\right). \qquad [17.60]$$

From [17.49]

$$\frac{1}{2\pi i}\int_c e^t\, t^{\alpha-\gamma}\mathrm{d}t = \frac{1}{\Gamma(\gamma - \alpha)} \qquad [17.61]$$

and with [17.45],

$$\frac{1}{\Gamma(\gamma - \alpha - 1)} = \frac{1}{\Gamma(\gamma - \alpha)}(\gamma - \alpha - 1),$$

we get

$$F_2 = \frac{\Gamma(\gamma)}{\Gamma(\gamma - \alpha)}(-x)^{-\alpha}\left\{1 - \frac{\alpha(\alpha - \gamma + 1)}{x} + ...\right\}. \qquad [17.62]$$

However, at one point in this derivation we "swindled," because the condition $|t/x| < 1$ for the convergence of the expansion [17.60] is only satisfied over one part of the path of integration (see Fig. 17.3). Therefore, we have to deal here with a so-called *asymptotic series*.[5] We must not take too many terms in such a series; otherwise the approximation will become poor again. For $\gamma - \alpha$ negative and integral, the result $F_2 = 0$ also follows from the asymptotic formula [17.62].

The expansion is completely analogous for F_1. We need only make the substitution

$$t - x = \tau$$

---

[5] For more information on this subject see, for example, E. T. WHITTAKER and G. N. WATSON, *A Course of Modern Analysis* (Cambridge University Press, New York, 1962).

to obtain

$$F_1(\alpha, \gamma, x) = \frac{\Gamma(\gamma)}{2\pi i} e^x \int_{C_2} e^\tau (x + \tau)^{\alpha-\gamma} \tau^{-\alpha} d\tau . \qquad [17.63]$$

This integral is formally the same as the one in [17.59] if we substitute $(\alpha - \gamma)$ for $-\alpha$ and $-x$ for $+x$ in the integrand. With these substitutions we can immediately write down the solution from [17.62]:

$$F_1(\alpha, \gamma, x) = \frac{\Gamma(\gamma)}{\Gamma(\alpha)} e^x x^{\alpha-\gamma} \left\{ 1 + \frac{(1-\alpha)(\gamma-\alpha)}{x} + ... \right\}. \qquad [17.64]$$

Of course, this too is only an asymptotic series. We again find $F_1 = 0$ for the case of the Laguerre polynomials, $-\alpha = k = 0, 1, 2, ...$

With the asymptotic formula

$$F(\alpha, \gamma, x) = F_1 + F_2 = \frac{\Gamma(\gamma)}{\Gamma(\gamma - \alpha)} (-x)^{-\alpha}$$
$$+ \frac{\Gamma(\gamma)}{\Gamma(\alpha)} e^x x^{\alpha-\gamma} + ... , \qquad [17.65]$$

and with

$$\gamma = m + 1, \qquad -\alpha = k , \qquad [17.66]$$

we can now write the solution [17.27] for the plane harmonic oscillator,

$$v_{k,m} = x^{m/2} e^{-x/2} F(-k, m+1, x) . \qquad [17.67]$$

For large x,

$$v_{k,m} \cong e^{+i\pi k} \frac{m!}{\Gamma(m+1+k)} x^{m/2+k} e^{-x/2}$$
$$+ \frac{m!}{\Gamma(-k)} x^{-(m/2+k+1)} e^{+x/2} . \qquad [17.68]$$

In the solution [17.67], x is real (see [17.15]); from this it follows that the second term in [17.68] diverges for large x. It is for this reason that the only allowed eigenvalues for a physically possible (that is, orthogonal and normalizable) solution are $k = 0, 1, 2, \ldots$. The asymptotic solution of the plane harmonic oscillator is then

$$v_{k,m} \cong e^{+i\pi k}\, \frac{m!}{\Gamma(m+1+k)}\, x^{m/2+k}\, e^{-x/2} . \qquad [17.69]$$

18. THE HYDROGEN ATOM

a. Separation of the wave equation in spherical coordinates

We now consider the Schrödinger equation for a central force field with potential $V(r)$,

$$-\frac{h^2}{2m}\, \nabla^2 u + V(r) \cdot u = E \cdot u , \qquad [18.1]$$

written in spherical coordinates r, ϑ, φ, which are related to the Cartesian coordinates x, y, z (Fig. 18.1) by the

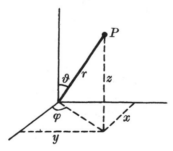

Figure 18.1

well-known formulas

$$\left.\begin{aligned} x &= r \sin\vartheta \cos\varphi \\ y &= r \sin\vartheta \sin\varphi \\ z &= r \cos\vartheta \end{aligned}\right\} . \qquad [18.2]$$

Then the Laplacian operator takes the form

$$\nabla^2 u = \frac{1}{r}\, \frac{\partial^2 (ru)}{\partial r^2} + \frac{1}{r^2}\left\{ \frac{1}{\sin\vartheta}\, \frac{\partial}{\partial\vartheta}\left(\sin\vartheta\, \frac{\partial u}{\partial\vartheta}\right) + \frac{1}{\sin^2\vartheta}\, \frac{\partial^2 u}{\partial\varphi^2} \right\} \qquad [18.3]$$

or

$$\nabla^2 u = \frac{1}{r^2} \frac{\partial}{\partial r}\left(r^2 \frac{\partial u}{\partial r}\right) + \frac{1}{r^2}\{\dots\}.$$

By assuming the form

$$u = v(r) \cdot Y(\vartheta, \varphi) \qquad [18.4]$$

for u, Eq. [18.1] splits into a radial part and an angle-dependent part:

$$\frac{1}{v}\frac{d}{dr}\left(r^2 \frac{dv}{dr}\right) + \frac{2mr^2}{h^2}\{E - V(r)\}$$
$$= -\frac{1}{Y}\left\{\frac{1}{\sin\vartheta}\frac{\partial}{\partial\vartheta}\left(\sin\vartheta \frac{\partial Y}{\partial\vartheta}\right) + \frac{1}{\sin^2\vartheta}\frac{\partial^2 Y}{\partial\varphi^2}\right\}. \qquad [18.5]$$

Since the left side of this equation depends only on r and the right side depends only on ϑ and φ, the two sides can be equated to a constant λ and separated:

$$\frac{1}{r^2}\frac{d}{dr}\left(r^2 \frac{dv}{dr}\right) + \left\{\frac{2m}{h^2}[E - V(r)] - \frac{\lambda}{r^2}\right\}v = 0, \qquad [18.6]$$

$$\frac{1}{\sin\vartheta}\frac{\partial}{\partial\vartheta}\left(\sin\vartheta \frac{\partial Y}{\partial\vartheta}\right) + \frac{1}{\sin^2\vartheta}\frac{\partial^2 Y}{\partial\varphi^2} + \lambda Y = 0. \qquad [18.7]$$

The only physically useful solutions of Eq. [18.7] are for

$$\lambda = l(l+1), \qquad l = 0, 1, 2, \dots; \qquad [18.8]$$

they are the *spherical harmonics* $Y_l(\vartheta, \varphi)$. Here, we content ourselves with writing down the most important properties of these functions, since we shall be concerned with them in more detail in an exercise (see Sec. 43).

b. The spherical harmonics

With condition [18.8] we now write the angle-dependent differential equation [18.7] as

$$\frac{1}{\sin\vartheta}\frac{\partial}{\partial\vartheta}\left(\sin\vartheta \frac{\partial Y}{\partial\vartheta}\right) + \frac{1}{\sin^2\vartheta}\frac{\partial^2 Y}{\partial\varphi^2} + l(l+1)Y = 0. \qquad [18.9]$$

This equation follows from the requirement of a central force field; however, it is independent of the particular form of the potential $V(r)$. We solve it with the trial solution

$$Y(\vartheta, \varphi) = \theta(\vartheta) e^{im\varphi}, \qquad [18.10]$$

in which the requirement of being single-valued leads to integral values of m. In this way we obtain the differential equation

$$\frac{1}{\sin\vartheta} \frac{d}{d\vartheta} \left(\sin\vartheta \frac{d\theta}{d\vartheta} \right) + \left\{ l(l+1) - \frac{m^2}{\sin^2\vartheta} \right\} \theta = 0 \,. \qquad [18.11]$$

With the substitutions

$$\left. \begin{array}{l} x = \cos\vartheta \\ \theta(\vartheta) = y(x) \end{array} \right\}, \quad \text{whence} \quad \left. \begin{array}{l} \sin\vartheta \, d\vartheta = - \, dx \\ \sin\vartheta \dfrac{d\theta}{d\vartheta} = - (1-x^2)y' \end{array} \right\}, \qquad [18.12]$$

the equation is transformed to

$$(1-x^2)y'' - 2xy' + \left\{ l(l+1) - \frac{m^2}{1-x^2} \right\} y = 0 \,. \qquad [18.13]$$

The solution of this equation can be written in the form

$$P_l^m(x) = (1-x^2)^{m/2} \frac{1}{2^l \, l!} \frac{d^{l+m}}{dx^{l+m}} (x^2-1)^l, \qquad [18.14]$$

where the condition

$$-l \leqslant m \leqslant +l \qquad [18.15]$$

on the integer m must be satisfied. The formula is normalized such that

$$P_l^0(1) = 1 \,. \qquad [18.16]$$

By applying formula [18.14] to positive as well as negative m, we obtain the relation

$$P_l^m = c_{lm} P_l^{-m}, \qquad c_{lm} = (-1)^m \frac{(l+m)!}{(l-m)!} \,. \qquad [18.17]$$

Therefore, the solution of the original differential equation [18.9] is

$$Y_{l,m}(\vartheta, \varphi) = P_l^m(\cos\vartheta)\, e^{im\varphi}\,. \qquad [18.18]$$

The functions $Y_{l,m}$ (or the P_l^m alone) are called *tesseral* or *spherical harmonics*. The nodal lines of their real and imaginary parts divide the surface of a sphere into four-sided regions (tesserae) of different signs, separated by parallels of latitude and meridians of longitude. The functions $Y_{l,0} = P_l^0$, written simply as Y_l and P_l, are called *zonal harmonics*, because they divide the sphere into latitudinal zones of different signs by their nodal lines. With the normalization [18.16], the $P_l(x)$ are the Legendre polynomials:

$$P_0 = 1, \quad P_1 = x, \quad P_2 = \tfrac{3}{2}x^2 - \tfrac{1}{2}, \quad P_3 = \tfrac{5}{2}x^3 - \tfrac{3}{2}x,$$
$$P_4 = \tfrac{35}{8}x^4 - \tfrac{15}{4}x^2 + \tfrac{3}{8}, \dots.$$

The spherical harmonic $Y_l(\vartheta, \varphi)$ is $1/r^l$ times a homogeneous polynomial of the lth degree in x, y, z which satisfies Laplace's equation:

$$Y_l(\vartheta, \varphi) = \frac{H_l(x, y, z)}{r^l}, \qquad \nabla^2 H_l = 0\,. \qquad [18.19]$$

The $Y_l(\vartheta, \varphi)$ can be expanded in terms of the linearly independent and orthogonal functions $Y_{l,m}(\vartheta, \varphi)$:

$$Y_l(\vartheta, \varphi) = \sum_{m=-l}^{+l} b_m Y_{l,m}(\vartheta, \varphi)\,. \qquad [18.20]$$

The value of the normalization integral is

$$N_l^m = \int_{-1}^{+1} \{P_l^m(x)\}^2\, \mathrm{d}x = \frac{(l+m)!}{(l-m)!}\frac{2}{2l+1}\,. \qquad [18.21]$$

Using this value we obtain

$$\overline{Y}_{l,m}(\vartheta, \varphi) = \frac{P_l^m(\cos\vartheta)}{\sqrt{N_l^m}}\frac{\exp[im\varphi]}{\sqrt{2\pi}} \qquad [18.22]$$

for the normalized spherical harmonics.

c. Solution of the radial differential equation

With the condition [18.8], Eq. [18.6] becomes

$$\frac{d^2}{dr^2}(rv) + \frac{2m}{h^2}\left\{E - V(r) - \frac{l(l+1)h^2}{2mr^2}\right\}rv = 0 . \qquad [18.23]$$

The eigenfunctions v satisfy the orthogonality relation

$$\int_0^\infty v_{E'}^* v_E r^2 \, dr = 0 \qquad \text{for} \qquad E' \neq E, \quad l' = l .$$

We see from the differential equation that there is formally an additional potential

$$l(l+1)h^2/2mr^2$$

for the radial motion in a central field. This expression is the analogue of the potential of the classical centrifugal force,

$$\overline{V}(r) = \frac{P^2}{2mr^2} \qquad (P = \text{angular momentum})$$

if we set

$$P^2 = l(l+1)h^2 .$$

This gives us an important relation between the quantum number l and the angular momentum $|P|$ of the particle.

In order for the radial differential equation [18.23] to be solved, the potential $V(r)$ must be given explicitly. We choose as an example the Coulomb potential:

$$V(r) = \frac{e_1 e_2}{r} , \qquad [18.24]$$

with

$e_1 = + Ze$ ($Z =$ atomic number),

$e_2 = -e$ ($e =$ magnitude of the elementary charge),

and obtain

$$\frac{d^2}{dr^2}(rv) + \frac{2m}{h^2}\left(E + \frac{Ze^2}{r}\right)(rv) - \frac{l(l+1)}{r^2}(rv) = 0. \qquad [18.25]$$

Our goal is the wave mechanical description of the system consisting of an electron plus a nucleus. For this purpose we should use the corresponding Schrödinger equation with the 6 coordinates x_1, y_1, z_1, x_2, y_2, z_2 of the two particles. Since the potential depends only on the relative coordinates, $V = V(x_1-x_2, y_1-y_2, z_1-z_2)$, this Schrödinger equation (in analogy to classical mechanics) can be split into an equation which describes the motion of the center of mass plus an equation which determines the relative motion of the two particles. The latter equation is formally the same as the equation for the motion of a particle of mass m moving in a potential V, where m is the reduced mass,

$$\frac{1}{m} = \frac{1}{m_{electron}} + \frac{1}{m_{nucleus}}. \qquad [18.26]$$

By substituting the reduced mass m (which is practically equal to $m_{electron}$ in this case) into [18.25] and solving this equation, the energy eigenvalues and the eigenfunctions for the relative motion of the electron and nucleus can be determined.

For this purpose we introduce the dimensionless quantities ϱ and ε:

$$r = \varrho a_0, \qquad E = \varepsilon E_0, \qquad [18.27]$$

$$a_0 = \frac{h^2}{Ze^2m} \qquad \text{(Bohr radius)}$$

$$E_0 = \frac{Z^2e^4m}{2h^2} = \frac{h^2}{2ma_0^2} \quad \begin{array}{l}\text{(energy corre-}\\\text{sponding to } a_0)\end{array} \qquad [18.28]$$

In terms of these quantities we obtain

$$\frac{d^2}{d\varrho^2}(\varrho v) + \left(\varepsilon + \frac{2}{\varrho} - \frac{l(l+1)}{\varrho^2}\right)\varrho v = 0. \qquad [18.29]$$

For small ϱ, we can immediately verify the solutions

$$v \sim \varrho^l, \qquad v \sim \varrho^{-l-1}; \qquad\qquad [18.30]$$

for large ϱ, by dropping terms of orders $1/\varrho$ and $1/\varrho^2$, we obtain

$$v \sim \frac{\exp[\pm \varrho \sqrt{-\varepsilon}]}{\varrho}. \qquad\qquad [18.31]$$

One way of solving the differential equation [18.29] is by using a power series as a trial solution; however, we want to profit from our knowledge of the hypergeometric functions. Taking account of the partial solutions [18.30] and [18.31], we assume

$$v = \varrho^l \exp[\pm \varrho \sqrt{-\varepsilon}] \cdot \omega(\varrho) \qquad\qquad [18.32]$$

as a trial solution, where $\omega(\varrho)$ is to remain finite for $\varrho \to 0$. If we compare the differential equation for ω,

$$\frac{\mathrm{d}^2\omega}{\mathrm{d}\varrho^2} + \left(\pm 2\sqrt{-\varepsilon} + \frac{2(l+1)}{\varrho} \right) \frac{\mathrm{d}\omega}{\mathrm{d}\varrho}$$

$$+ \frac{2 \pm 2(l+1)\sqrt{-\varepsilon}}{\varrho} \omega = 0 \qquad [18.33]$$

(first dividing by -4ε) with the equation for the hypergeometric function $F(\alpha, \gamma, x)$ (see [17.51]),

$$\frac{\mathrm{d}^2 F}{\mathrm{d}x^2} + \left(\frac{\gamma}{x} - 1 \right) \frac{\mathrm{d}F}{\mathrm{d}x} - \frac{\alpha}{x} F = 0, \qquad\qquad [18.34]$$

we see that they are identical if we put

$$\alpha = \pm \frac{1}{\sqrt{-\varepsilon}} + l + 1, \qquad\qquad [18.35]$$

$$\gamma = 2(l+1), \qquad\qquad [18.36]$$

$$x = \mp 2\varrho\sqrt{-\varepsilon}; \qquad\qquad [18.37]$$

thus, ω is given by

$\omega = \text{constant}$

$$\times F\left(\pm\frac{1}{\sqrt{-\varepsilon}}+l+1,\ 2(l+1),\ \mp 2\varrho\sqrt{-\varepsilon}\right). \qquad [18.38]$$

From the power series expansion [17.54],

$$F(\alpha,\gamma,x)=1+\frac{\alpha}{\gamma}x+\cdots,$$

it follows that the solution [18.32] actually does begin with ϱ^l for small ϱ, as we conjectured in [18.30].

In order to carry the discussion further, we must distinguish between $\varepsilon < 0$ and $\varepsilon > 0$, that is, between $\sqrt{-\varepsilon}$ real and imaginary.

d. Case 1: $\varepsilon < 0$. The discrete energy spectrum

We take the lower sign for $\sqrt{-\varepsilon}$ throughout; in [18.32] this means that we take the minus sign. The upper sign does not lead to a physically acceptable solution, because the normalization integral diverges.

Using [18.35], [18.36], [18.37], and the *radial quantum number* n_r,

$$n_r = -\alpha = \frac{1}{\sqrt{-\varepsilon}}-l-1, \qquad [18.39]$$

then for large ϱ we can write the asymptotic formula [17.65] for $F(\alpha,\gamma,x)$ as

$$F\sim\Gamma(2l+2)\left\{\frac{1}{\Gamma[(1/\sqrt{-\varepsilon})+l+1]}(-2\varrho\sqrt{-\varepsilon})^{r_r}\right.$$

$$\left.+\frac{1}{\Gamma(-n_r)}\exp[2\varrho\sqrt{-\varepsilon}](+2\varrho\sqrt{-\varepsilon})^{-(1/\sqrt{-\varepsilon})-l-1}\right\}. \qquad [18.40]$$

Because of the second term, this solution approaches infinity exponentially for $\varrho \to \infty$; such behavior is incompatible with the physics of the problem. We must demand that this term vanish; this means that we must require

$$n_r = 0, 1, 2, \ldots . \qquad [18.41]$$

We then obtain the asymptotic expression

$$v \sim \exp[-\varrho\sqrt{-\varepsilon}] \cdot \varrho^{(1/\sqrt{-\varepsilon})-1} \qquad [18.42]$$

for the solution [18.32]. According to formula [17.57], the condition [18.41] reduces the hypergeometric function $F(\alpha, \gamma, x)$ to the simpler associated Laguerre polynomial,

$$L_{m+k}^{m} = \text{constant} \times F(-k, m+1, x) . \qquad [18.43]$$

Therefore, according to the trial solution [18.32], the eigenfunctions of the radial differential equation [18.29] are

$$v = \text{constant} \times \varrho^l \exp\left[-\frac{\varrho}{n_r+l+1}\right] L_{2l+1+n_r}^{2l+1}\left(\frac{2}{n_r+l+1}\varrho\right).$$

The normalization can be carried out in a way analogous to that for the plane harmonic oscillator.

1. *The quantum numbers. Energy eigenvalues and degeneracy.* The *principal quantum number* n is defined by

$$n = \frac{1}{\sqrt{-\varepsilon}} = n_r + l + 1 . \qquad [18.44]$$

Its values are determined by the condition [18.41]:

$$n = l+1, l+2, \ldots . \qquad [18.45]$$

These relations also imply that for a given n the *angular momentum quantum number* l can only take the values

$$l = 0, 1, \ldots, n-1 . \qquad [18.46]$$

The energy eigenvalues follow from [18.44] and [18.28]:

$$E_n = \varepsilon E_0 = -\frac{1}{n^2}\frac{Z^2 e^4 m}{2h^2} . \qquad [18.47]$$

With these results we can write the famous *Balmer formula*:

$$hv = E' - E'', \quad \text{or} \quad v = R\left(\frac{1}{n''^2} - \frac{1}{n'^2}\right), \qquad R = \frac{Z^2 e^4 m}{2h^3}.$$

It is worth noting that the eigenvalues E_n do not depend on the angular momentum quantum number l at all. For a fixed l, we have seen that there exist $2l+1$ different spherical harmonics or eigenfunctions of the Schrödinger equation [18.1]. To this $(2l+1)$-fold degeneracy, which is present for a completely arbitrary central field $V(r)$, there must be added the n-fold degeneracy [18.46], which is related to the Coulomb field. Thus, for the total degree of degeneracy g_n of a state with principal quantum number n, we find

$$g_n = \sum_{l=0}^{n-1} (2l+1) = \sum_{l=0}^{n-1} \{(l+1)^2 - l^2\} = n^2. \qquad [18.48]$$

We did not find this degree of degeneracy in any of the cases that we solved previously.

We note that the grounds tate $(n=1, l=0)$ is nondegenerate. Furthermore, since $\varepsilon = -1$, the solution reduces in this case to

$$u = \text{constant} \times e^{-\varrho}. \qquad [18.49]$$

2. *Examples in which the degeneracy is removed.* The valence electron of an alkali atom moves in a central field which, however, is not quite the same as the Coulomb field. This has as a consequence that the n-fold degeneracy due to the Coulomb field is removed, and that the nth hydrogen-like level of the valence electron splits into n levels. We have a further example if there is an external magnetic field. The magnetic field alters the character of the original central field completely, with the result that the nth level splits into n^2 levels.

e. *Case 2:* $\varepsilon > 0$. *The continuous energy spectrum*

We come to the case which describes a "free" particle. With [18.38] and the assumption

$$\sqrt{-\varepsilon} = +i\sqrt{\varepsilon} \qquad \text{(positive root!)}, \qquad [18.50]$$

we can write the solution [18.32] as

$$v(\varrho) = \text{constant}$$
$$\times \varrho^l \exp[-i\sqrt{\varepsilon}\varrho] F\left(\frac{i}{\sqrt{\varepsilon}} + l + 1,\, 2l + 2,\, 2i\varrho\sqrt{\varepsilon}\right), \qquad [18.51]$$

where we again restrict ourselves to the "lower" sign of $\pm\sqrt{\varepsilon}$. Introduction of the wave number k of the free particle,

$$kh = \sqrt{2mE}, \qquad [18.52]$$

leads to the relations (see [18.27])

$$\left.\begin{array}{l} ka_0 = \sqrt{\varepsilon} = \sqrt{\dfrac{E}{E_0}} \\[2ex] \sqrt{\varepsilon}\varrho = \dfrac{\sqrt{\varepsilon}}{a_0} r = kr \qquad \left(a_0 = \dfrac{h^2}{Ze^2m}\right) \end{array}\right\}, \qquad [18.53]$$

with which the solution [18.51] can be written in the somewhat clearer form

$$v(r) = \text{constant}$$
$$\times (kr)^l\, e^{-ikr} F\left(\frac{i}{ka_0} + l + 1,\, 2l + 2,\, 2ikr\right). \qquad [18.54]$$

As a matter of fact, this solution is actually real. To prove this, we separate F into two parts,

$$F = F_1 + F_2, \qquad [18.55]$$

and use the integral representation [17.59],

$$F_{\frac{1}{2}} = \frac{\Gamma(2l + 2)}{2\pi i}$$
$$\times \int_{\substack{C_1 \\ C_2}} e^t t^{(i/ka_0)-l-1}(t - 2ikr)^{-(i/ka_0)-l-1}\mathrm{d}t, \qquad [18.56]$$

for $F_{\frac{1}{2}}$. If, for example, we substitute

$$t \to t' - 2ikr \qquad [18.57]$$

in the expression for F_2, it follows that the path of integration C_2 is the same as C_1, from which

$$e^{-ikr}F_2 = (e^{-ikr}F_1)^* \qquad [18.58]$$

follows immediately. Thus, we can write

$$e^{ikr}F = e^{-ikr}F_1 + (e^{-ikr}F_1)^*,$$

which proves the assertion.

Now, we want to investigate the asymptotic behavior of the solution [18.54]. Using [18.58], the asymptotic formula for F_1, [17.64], and the relations [18.35] to [18.37] and [18.53], the solution for large r can be written in the form

$$v(r) = \text{constant} \times (kr)^l\, e^{-ikr} F \sim \text{constant}$$
$$\times \frac{\Gamma(2l+2)}{\Gamma[(i/ka_0)+l+1]} (kr)^l\, e^{+ikr} (2ikr)^{(i/ka_0)-l-1}$$
$$+ \text{complex conjugate}. \qquad [18.59]$$

Using

$$2ikr = \exp\left[\log(2kr) + i\frac{\pi}{2}\right]$$

and

$$(2ikr)^{(i/ka_0)-l-1} = 2^{-l-1}(kr)^{-l-1}\exp\left[\frac{i}{ka_0}\log(2kr)\right]$$
$$\times \exp\left[+i\frac{\pi}{2}\left(\frac{i}{ka_0}-l-1\right)\right],$$

we can also write this expression in the form

$$v(r) = \text{constant} \times \frac{(2l+1)!}{\Gamma[(i/ka_0)+l+1]}$$
$$\times \frac{1}{kr}\exp\left[ikr + \frac{i}{ka_0}\log(2kr) - i\frac{\pi}{2}(l+1)\right]$$
$$\times \frac{1}{2^{l+1}}\exp\left[-\frac{\pi}{2ka_0}\right] + \text{complex conjugate}. \qquad [18.60]$$

With the relation

$$\Gamma\left(\frac{i}{ka_0} + l + 1\right) = |\Gamma|\exp[i\sigma(l, ka_0)] \qquad [18.61]$$

(where σ is the phase of the Γ function), we finally obtain for $kr \gg 1$

$$v_l(r) = \text{constant} \times \frac{(2l+1)!}{2^l} \frac{\exp[-\pi/2ka_0]}{|\Gamma[(i/ka_0) + l + 1]|} \frac{1}{kr}$$

$$\times \cos\left\{kr + \frac{1}{ka_0}\log(2kr) - \frac{\pi}{2}(l+1) - \sigma(l, ka_0)\right\}. \qquad [18.62]$$

Essentially, this yields a spherical wave for the solution, which, however, contains a logarithmic correction to the phase. The effect of the correction is to make the phase of the wave function for large r change more rapidly than kr. This correction, which is typical of the Coulomb field, can also be justified classically.

For $Z = 0$, that is, for $1/a_0 = 0$ (see [18.28]), the force-free solution follows from [18.62] (see also [21.10]).

We shall show in an exercise (see Sec. 46) that normalization according to

$$\int_0^\infty v_{l,k} v_{l,k'} r^2 \, \mathrm{d}r = \delta(k' - k) \qquad [18.63]$$

leads to the normalized asymptotic formula

$$v_{l,k}(r) = \sqrt{\frac{2}{\pi}} \frac{1}{r}$$

$$\times \sin\left\{kr + \frac{1}{ka_0}\log(2kr) - \frac{\pi}{2}l - \sigma(l, ka_0)\right\}. \qquad [18.64]$$

Remark: As is evident from the above formulas, the states of positive energy ($\varepsilon > 0$) are infinitely degenerate. This is because, for a given value of k, the quantum number l can independently assume all values between 0 and ∞

f. Solution of the wave equation in parabolic coordinates

The wave equation for the hydrogen atom can also be separated and solved in parabolic coordinates.[6] We want to obtain this solution and show, as is to be expected on physical grounds, that the same energy eigenvalues and the same degeneracy result as with the spherical coordinates. Further, we shall also treat the scattering problem with this solution and derive a solution corresponding to formula [18.62] whose characteristics are more evident.

The connection between the parabolic coordinates λ_1, λ_2,

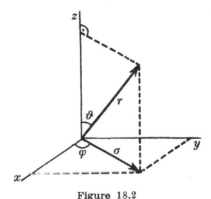

Figure 18.2

φ and the spherical polar coordinates (Fig. 18.2) is given by

$$\left.\begin{aligned} \lambda_1 &= r + z = r(1 + \cos\vartheta) \\ \lambda_2 &= r - z = r(1 - \cos\vartheta) \\ \varphi &= \varphi \end{aligned}\right\} . \qquad [18.65]$$

In addition, we have

$$\left.\begin{aligned} r &= \tfrac{1}{2}(\lambda_1 + \lambda_2) \\ z &= \tfrac{1}{2}(\lambda_1 - \lambda_2) \\ \sigma^2 &= r^2 - z^2 = \lambda_1\lambda_2 \end{aligned}\right\} . \qquad [18.66]$$

[6] See A. SOMMERFELD, "Über die Beugung und Bremsung von Elektronen," *Ann. Physik* **11**, 268 (1931).

This time we immediately write the wave equation [18.1] with the Coulomb potential [18.24], using the dimensionless coordinates $\varepsilon = E/E_0$ and $\varrho = r/a_0$ given by [18.27]:

$$\nabla^2 u + \left(\varepsilon + \frac{2}{\varrho}\right)u = 0. \qquad [18.67]$$

In the following, until the treatment of the continuous spectrum, we also use dimensionless parabolic coordinates; these follow from formulas [18.65] and [18.66] by making the substitutions $r \to r/a_0$, $z \to z/a_0$. For the sake of simplicity, we also call these dimensionless quantities λ_1 and λ_2.

g. Separation of the wave equation [18.67] in parabolic coordinates

In this case the square of the element of length is (dimensionless quantities!)

$$ds^2 = d\sigma^2 + dz^2 + \sigma^2 d\varphi^2$$

$$= \frac{1}{4\lambda_1\lambda_2}(\lambda_1 d\lambda_2 + \lambda_2 d\lambda_1)^2 + \frac{1}{4}(d\lambda_1 - d\lambda_2)^2 + \lambda_1\lambda_2 d\varphi^2$$

$$= \frac{\lambda_1 + \lambda_2}{4}\left\{\frac{d\lambda_1^2}{\lambda_1} + \frac{d\lambda_2^2}{\lambda_2} + 4\left(\frac{1}{\lambda_1} + \frac{1}{\lambda_2}\right)^{-1}d\varphi^2\right\}. \qquad [18.68]$$

As it must be with orthogonal coordinates, there are no mixed terms. It can be shown[7] that the Laplacian operator in orthogonal, curvilinear coordinates x_1, x_2, x_3 is of the general form

$$\nabla^2 u = \frac{1}{e_1 e_2 e_3}\frac{\partial}{\partial x_1}\left(\frac{e_2 e_3}{e_1}\frac{\partial u}{\partial x_1}\right) + \frac{1}{e_1 e_2 e_3}\frac{\partial}{\partial x_2}\left(\frac{e_3 e_1}{e_2}\frac{\partial u}{\partial x_2}\right)$$

$$+ \frac{1}{e_1 e_2 e_3}\frac{\partial}{\partial x_3}\left(\frac{e_1 e_2}{e_3}\frac{\partial u}{\partial x_3}\right), \qquad [18.69]$$

with

$$ds^2 = \sum_{k=1}^{3} e_k^2 dx_k^2, \qquad dV = e_1 e_2 e_3 dx_1 dx_2 dx_3. \qquad [18.70]$$

---

[7] See W. PAULI, *Lectures in Physics: Electrodynamics* (M.I.T. Press, Cambridge, Mass., 1972).

In our case this leads to

$$\nabla^2 u = \frac{4}{\lambda_1 + \lambda_2} \left\{ \frac{\partial}{\partial \lambda_1} \left(\lambda_1 \frac{\partial u}{\partial \lambda_1} \right) + \frac{\partial}{\partial \lambda_2} \left(\lambda_2 \frac{\partial u}{\partial \lambda_2} \right) \right.$$

$$\left. + \frac{1}{4} \left(\frac{1}{\lambda_1} + \frac{1}{\lambda_2} \right) \frac{\partial^2 u}{\partial \varphi^2} \right\}. \qquad [18.71]$$

With this result we can write the wave equation [18.67] in parabolic coordinates as

$$\frac{\partial}{\partial \lambda_1} \left(\lambda_1 \frac{\partial u}{\partial \lambda_1} \right) + \frac{\partial}{\partial \lambda_2} \left(\lambda_2 \frac{\partial u}{\partial \lambda_2} \right) + \frac{1}{4} \left(\frac{1}{\lambda_1} + \frac{1}{\lambda_2} \right) \frac{\partial^2 u}{\partial \varphi^2}$$

$$+ \left\{ \frac{\varepsilon}{4} (\lambda_1 + \lambda_2) + 1 \right\} u = 0. \qquad [18.72]$$

This equation can be separated with the trial solution

$$u = f_1(\lambda_1) \cdot f_2(\lambda_2) \cdot e^{\pm i m \varphi}; \qquad [18.73]$$

the result is

$$\frac{d}{d\lambda_1} \left(\lambda_1 \frac{\partial f_1}{\partial \lambda_1} \right) + \left(-\frac{1}{4} \frac{m^2}{\lambda_1} + \frac{\varepsilon}{4} \lambda_1 + \frac{1+\beta}{2} \right) f_1 = 0, \qquad [18.74]$$

$$\frac{d}{d\lambda_2} \left(\lambda_2 \frac{\partial f_2}{\partial \lambda_2} \right) + \left(-\frac{1}{4} \frac{m^2}{\lambda_2} + \frac{c}{4} \lambda_2 + \frac{1-\beta}{2} \right) f_2 = 0, \qquad [18.75]$$

where the parameter β is the arbitrary separation constant. These two equations have the form of Eq. [17.16] for the plane harmonic oscillator, which we now write as

$$(xf')' + \left\{ -\frac{x}{4} + \left(n_i + \frac{m+1}{2} \right) - \frac{m^2}{4x} \right\} f = 0, \quad i = 1, 2 \qquad [18.76]$$

$(\lambda/4 \to n_i + (m+1)/2, \ k \to n_i)$. We found

$$f = \text{constant} \times x^{m/2} \, e^{-x/2} L_{m+n_i}^m (x)$$

$$= \text{constant} \times x^{m/2} \, e^{-x/2} F(-n_i, m+1, x) \qquad [18.77]$$

to be the only useful solution of this equation (see [17.27]). The numbers n_1, n_2, and m are quantum numbers associated with the parabolic coordinates λ_1, λ_2, and φ.

h. Discrete energy spectrum ($\varepsilon < 0$)

We can bring [18.76] into agreement with [18.74] and [18.75] by setting

$$x = \lambda_i \sqrt{-\varepsilon} \qquad (\sqrt{-\varepsilon} > 0) \qquad [18.78]$$

and choosing β such that

$$\left.\begin{array}{l} \dfrac{1}{2}(1+\beta) = \sqrt{-\varepsilon}\left(n_1 + \dfrac{m+1}{2}\right) \\[2mm] \dfrac{1}{2}(1-\beta) = \sqrt{-\varepsilon}\left(n_2 + \dfrac{m+1}{2}\right) \quad (n_i = 0, 1, 2, \ldots) \end{array}\right\} . \quad [18.79]$$

From this,

$$1 = \sqrt{-\varepsilon}(n_1 + n_2 + m + 1) \qquad [18.80]$$

follows immediately. With

$$n = n_1 + n_2 + m + 1 \qquad (n = 1, 2, \ldots) \qquad [18.81]$$

this can also be written in the form

$$\varepsilon = -\frac{1}{n^2}, \qquad [18.82]$$

which corresponds to the expression [18.47] already found for the energy eigenvalues. In addition we obtain

$$\beta = \sqrt{-\varepsilon}(n_1 - n_2) = \frac{n_1 - n_2}{n}. \qquad [18.83]$$

The solution [18.73] can now be written in detail ($x \to \lambda_i/n$):

$$u = \text{constant} \times \exp\left[-\frac{\lambda_1 + \lambda_2}{2n}\right]\left(\frac{\lambda_1 \lambda_2}{n^2}\right)^{m/2} L_{m+n_1}^m\left(\frac{\lambda_1}{n}\right)$$

$$\times L_{m+n_2}^m\left(\frac{\lambda_2}{n}\right) e^{\pm im\varphi}. \qquad [18.84]$$

Remark: The separation of the wave equation in parabolic coordinates is also possible when there is an external field F, which adds a term

$$\text{constant} \times (F \cdot z)$$

to the potential V in [18.1]. The only consequence of this is that the separated Eqs. [18.74] and [18.75] have additional terms

$$+\text{constant} \times \lambda_1^2 f_1 F \quad \text{and} \quad -\text{constant} \times \lambda_2^2 f_2' F,$$

respectively. If F is small, these equations can be solved approximately by means of the Born approximation (see Sec. 24). An interesting result is that the energy eigenvalues have the form

$$\varepsilon = -\frac{1}{n^2} + \text{constant} \times (n_1 - n_2) n F$$

in first approximation (first-order Stark effect).[8]

i. Continuous spectrum $(\varepsilon > 0)$

We now return to the usual units; that is, we use $a_0 \lambda_i'$ instead of λ_i. As with [18.52], we introduce the wave number k and, with the help of [18.53],

$$\sqrt{-\varepsilon} = \mp i\sqrt{\varepsilon} = \mp i k a_0 \qquad [18.85]$$

(the upper sign is for index 1), we write the condition [18.79] in the form

$$\left.\begin{array}{l} \dfrac{1}{2}(1+\beta) = -ika_0\left(v_1 + \dfrac{m+1}{2}\right) \\[2mm] \dfrac{1}{2}(1-\beta) = ika_0\left(v_2 + \dfrac{m+1}{2}\right) \end{array}\right\}. \qquad [18.86]$$

Addition of the two equations yields

$$v_1 - v_2 = \frac{i}{ka_0}, \qquad [18.87]$$

---

[8] See, for example, A. SOMMERFELD, *Atombau und Spektrallinien*, vol. 2.

where, in contrast to the n_i, the v_i are not integers. In this notation, the differential equation [18.72] is

$$\frac{\partial}{\partial \lambda_1}\left(\lambda_1 \frac{\partial u}{\partial \lambda_1}\right) + \frac{\partial}{\partial \lambda_2}\left(\lambda_2 \frac{\partial u}{\partial \lambda_2}\right) - \frac{m^2}{4}\left(\frac{1}{\lambda_1} + \frac{1}{\lambda_2}\right)u$$

$$+ \left\{\frac{k^2}{4}(\lambda_1 + \lambda_2) + \frac{1}{a_0}\right\}u = 0 , \qquad [18.88]$$

and its solution is $(x \to \mp ik\lambda_i)$

$$u = \text{constant} \times \exp\left[ik\,\frac{\lambda_1 - \lambda_2}{2}\right](k^2\lambda_1\lambda_2)^{m/2}F(\alpha, m+1, -ik\lambda_1)$$

$$\times F\left(\alpha + \frac{i}{ka_0}, m+1, ik\lambda_2\right)e^{\pm im\varphi} . \qquad [18.89]$$

In this formula we have replaced the still undetermined v_1 by $-\alpha$.

Remark: All of our formulas and expressions are written for positive a_0, that is, for attractive forces. In the case of repulsive forces (which only makes sense for $\varepsilon > 0$), we must make the substitution $a_0 \to -a_0$.

Chapter 6. Collision Processes

We concern ourselves here with the scattering of particles by particles, without taking spin interactions into consideration. The solution of this problem follows from the formulas which we obtained in the last chapter for the case of a continuous energy spectrum ($\varepsilon > 0$).

The starting point for the present considerations is the solution [18.89]. The incident particle current will be represented by a plane wave propagating in the positive z direction; of course, the wave will be symmetric about the z axis, and we can therefore restrict ourselves to the case $m = 0$. For the special case of $\alpha = 0$, we shall show that the solution [18.89],

$$u = \text{constant} \times \exp\left[\frac{ik}{2}(\lambda_1 - \lambda_2)\right] F\left(\frac{i}{ka_0}, 1, ik\lambda_2\right), \quad [\text{I}]$$

can be written asymptotically as an incident plane wave plus the scattered wave. For the general case of $\alpha \neq 0$, the solution contains terms which represent a plane wave propagating in the negative z direction and an incoming spherical wave; this follows immediately from the asymptotic formula [17.65] for the hypergeometric function F. Therefore, this case is incompatible with the physical problem as we have stated it.

Remark: In the relativistic case it is not possible to separate the wave equation in parabolic coordinates. In particular, this simple decomposition of the asymptotic solution is also not valid.

19. ASYMPTOTIC SOLUTION OF THE SCATTERING PROBLEM

The asymptotic formula [17.65] for the hypergeometric function $F(i/ka_0, 1, ik\lambda_2)$ is

$$F = \frac{1}{\Gamma(1-i/ka_0)}(-ik\lambda_2)^{-i/ka_0}\left\{1 + \frac{1}{ik\lambda_2(ka_0)^2}\right\}$$
$$+ \frac{1}{\Gamma(i/ka_0)}\exp[ik\lambda_2](ik\lambda_2)^{-1+i/ka_0}; \qquad [19.1]$$

in this expression we have used the first approximation given in [17.62]. Using [18.65] and [18.66], we again introduce polar coordinates r and ϑ:

$$\left.\begin{aligned}\lambda_1 &= r + z \\ \lambda_2 &= r - z = r(1-\cos\vartheta) = 2r\sin^2\tfrac{1}{2}\vartheta \\ \lambda_1 - \lambda_2 &= 2z\end{aligned}\right\}. \qquad [19.2]$$

Written out in detail, the solution [I] is now

$$\begin{aligned}u = c\,\frac{\exp[-\pi/2ka_0]}{\Gamma(1-i/ka_0)}&\left\{\exp\left[i\left(kz - \frac{1}{ka_0}\log(2kr\sin^2\tfrac{1}{2}\vartheta)\right)\right]\right.\\ &\times\left(1 - \frac{i}{2kr(ka_0)^2\sin^2\tfrac{1}{2}\vartheta}\right) - i\,\frac{\Gamma(1-i/ka_0)}{\Gamma(i/ka_0)}\\ &\left.\times\frac{\exp\left[i(kr + k^{-1}a_0^{-1}\log(2kr\sin^2\tfrac{1}{2}\vartheta))\right]}{2kr\sin^2\tfrac{1}{2}\vartheta}\right\}. \qquad [19.3]\end{aligned}$$

In obtaining this expression we have used

$$\left.\begin{aligned}(-i)^{-i/ka_0} &= \exp\left[-\frac{\pi}{2ka_0}\right] \\ (k\lambda_2)^{-i/ka_0} &= \exp\left[-\frac{i}{ka_0}\log(2kr\sin^2\tfrac{1}{2}\vartheta)\right]\end{aligned}\right\}. \qquad [19.4]$$

In order to further simplify the expression, we use the formula

$$\begin{aligned}\frac{\Gamma(1-i/ka_0)}{i\Gamma(i/ka_0)} &= \frac{1}{ka_0}\frac{\Gamma(1-i/ka_0)}{\Gamma(1+i/ka_0)} \\ &= \frac{1}{ka_0}\exp[-2i\sigma(0, ka_0)], \qquad [19.5]\end{aligned}$$

which follows straightforwardly from the functional equation

$$\Gamma\left(1+\frac{i}{ka_0}\right) = \frac{i}{ka_0}\Gamma\left(\frac{i}{ka_0}\right)$$ [19.6]

and the definition of $\sigma(l, ka_0)$ given in [18.61]. With the normalization

$$c \cdot \frac{\exp[-\pi/2ka_0]}{\Gamma(1-i/ka_0)} = 1 ,$$ [19.7]

the asymptotic solution [19.3] can now be written in the form

$$u = \exp\left[i\left(kz - \frac{1}{ka_0}\log(2kr\sin^2\tfrac{1}{2}\vartheta)\right)\right]\left(1 - \frac{i}{2k^3a_0^2r\sin^2\tfrac{1}{2}\vartheta}\right)$$
$$+ \frac{\exp\{i[kr + k^{-1}a_0^{-1}\log(2kr\sin^2\tfrac{1}{2}\vartheta) - 2\sigma(0, ka_0)]\}}{2k^2a_0r\sin^2\tfrac{1}{2}\vartheta} .$$ [19.8]

By assumption this formula is only valid for

$$k\lambda_2 - 2kr\sin^2\tfrac{1}{2}\vartheta \gg 1 ;$$ [19.9]

that is, it is valid in all of space except for a parabolic region defined by [19.9] (Fig. 19.1).

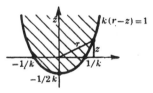

Figure 19.1

To lowest order, the solution [19.8] is of the form

$$u \sim \exp[ikz] + f(\vartheta)\frac{\exp[ikr]}{r} ;$$ [19.10]

that is, $u \sim$ incident plane wave + outgoing spherical wave. In the case of a Coulomb field, with which we have calculated so far, the following substitutions must be made for the phases:

$$kz \rightarrow kz - \frac{1}{ka_0} \log(2kr \sin^2 \tfrac{1}{2}\vartheta) \Bigg\rbrace$$
$$kr \rightarrow kr + \frac{1}{ka_0} \log(2kr \sin^2 \tfrac{1}{2}\vartheta) \Bigg.$$

[19.11]

This logarithmic phase correction does not occur if the potential falls off more rapidly,

$$\lim_{r \to \infty} V(r) \cdot r = 0 \,.$$

In an exercise (see Sec. 46) we shall note that another asymptotic representation of the wave function, one for a fixed value of l, is possible for such a potential:

$$u \sim \frac{\text{constant}}{r} \times \sin\left(kr - l\frac{\pi}{2} + \delta_l\right) \,.$$

[19.12]

In formula [18.62] we have already derived the analogous asymptotic expression for the Coulomb potential, which follows from [19.12] with the substitution [19.11].

20. THE SCATTERING CROSS SECTION. THE RUTHERFORD SCATTERING FORMULA

Using the solution given by [19.8] and [19.10], we can calculate the differential scattering cross section dQ. The definition of dQ is

$$\mathrm{d}Q = \frac{i_s}{i_0} = \frac{\left(\begin{array}{c}\text{number of particles scattered into}\\ \text{solid angle } \mathrm{d}\Omega \text{ per unit time}\end{array}\right)}{\left(\begin{array}{c}\text{number of incident particles per}\\ \text{unit area per unit time}\end{array}\right)} \,.$$

[20.1]

If the solution of the scattering problem is of the form [19.10], then we have

$$\mathrm{d}Q = |f(\vartheta)|^2 \mathrm{d}\Omega \,.$$

[20.2]

We prove this formula by calculating the ratio of the scattered-particle current to the incident-particle current. From the general formula for the particle current density [7.10],

$$i = \frac{h}{2mi}(\psi^* \operatorname{grad}\psi - \psi \operatorname{grad}\psi^*),$$

we obtain

$$i_0 = \frac{hk}{m} \qquad [20.3]$$

for the (incident) plane wave; for the wave scattered into solid angle $d\Omega$, we immediately obtain

$$i_s = \frac{hk}{mr^2}|f(\vartheta)|^2 r^2 d\Omega \qquad [20.4]$$

(remember $kr \gg 1$ and $\operatorname{grad} \sim d/dr$). Thus, formula [20.2] is proved.

From [20.2] we obtain

$$dQ = \frac{1}{4k^4 a_0^2 \sin^4 \frac{1}{2}\vartheta} d\Omega \qquad [20.5]$$

for the solution [19.8]. With

$$k^2 a_0 = \left(\frac{mv}{h}\right)^2 a_0 = \frac{2mE}{h^2}\frac{h^2}{Ze^2 m} = \frac{2E}{Ze^2}, \qquad [20.6]$$

this yields the famous Rutherford scattering formula,

$$dQ = \frac{Z^2 e^4}{16E^2 \sin^4 \frac{1}{2}\vartheta} d\Omega = \frac{Z^2 e^4}{4m^2 v^4 \sin^4 \frac{1}{2}\vartheta} d\Omega. \qquad [20.7]$$

This expression, as well as formula [19.8], can also be derived classically; [1] it is a result of the particular Coulomb potential. With other potentials it is generally not possible to obtain the same results classically and wave mechanically. [2]

[1] W. GORDON, Z. Physik 48, 188 (1928).
[2] Th. SEXL, Z. Physik 67, 766 (1931).

21. SOLUTION OF THE FORCE-FREE WAVE EQUATION

Now we want to try to solve the scattering problem directly, without reference to the general asymptotic formula [17.65]. As a result we shall understand the problem somewhat better.

Toward this end, we first solve the force-free wave equation

$$\nabla^2 u + k^2 u = 0 \qquad [21.1]$$

in polar coordinates. With the usual trial solution [18.4],

$$u = v_l(r)\, Y_l(\vartheta, \varphi)\,,$$

we separate this differential equation into an angle-dependent part and a part depending only on r. The part depending on r is $\big($see [18.6], with $V=0\big)$

$$\frac{1}{r}\frac{d^2}{dr^2}(rv_l) + \left(k^2 - \frac{l(l+1)}{r^2}\right)v_l = 0\,, \qquad [21.2]$$

or, written in terms of the dimensionless coordinate $\varrho = kr$,

$$\frac{1}{\varrho}\frac{d^2}{d\varrho^2}(\varrho v_l) + \left(1 - \frac{l(l+1)}{\varrho^2}\right)v_l = 0\,. \qquad [21.3]$$

By induction $(l \to l+1)$ we obtain

$$v_{l+1} = -\varrho^l \frac{d}{d\varrho}(\varrho^{-l} v_l) = -\frac{dv_l}{d\varrho} + \frac{l}{\varrho}v_l\,. \qquad [21.4]$$

Using this, we can write the solution of [21.3] in the form

$$v_l = \varrho^l(-1)^l \left(\frac{d}{\varrho\, d\varrho}\right)^l v_{l=0}\,. \qquad [21.5]$$

From [21.3] we immediately obtain two independent solutions for $v_{l=0}$:

$$\left.\begin{array}{ll} \xi_0^{(1)} = \dfrac{\exp[i\varrho]}{i\varrho} & \text{outgoing spherical wave} \\[2ex] \xi_0^{(2)} = -\dfrac{\exp[-i\varrho]}{i\varrho} & \text{incoming spherical wave} \end{array}\right\}\,. \qquad [21.6]$$

From these we obtain, for example, the solution

$$\psi_0 = \frac{1}{2}(\xi_0^{(1)} + \xi_0^{(2)}) = \frac{\sin\varrho}{\varrho}, \qquad [21.7]$$

which is the only solution regular at $\varrho = 0$. With [21.5] we can now write

$$\xi_l^{(1)}_{(2)} = \varrho^l(-1)^l\left(\frac{d}{\varrho \, d\varrho}\right)^l \xi_0^{(1)}_{(2)}, \qquad [21.8]$$

$$\left.\begin{array}{l} \psi_l = \frac{1}{2}(\xi_l^{(1)} + \xi_l^{(2)}) \\ \quad = \varrho^l(-1)^l\left(\dfrac{d}{\varrho \, d\varrho}\right)^l \psi_0 \\ \chi_l = (\xi_l^{(1)} - \xi_l^{(2)})/2i \end{array}\right\} \qquad [21.9]$$

We now want to write this solution in a somewhat simpler form for the two special cases $\varrho \gg 1$ and $\varrho \ll 1$. For $\varrho \gg 1$, all that we need differentiate in [21.8] is $e^{i\varrho}$, because the other terms are of higher order in $1/\varrho$:

$$\xi_l^{(1)} = (-1)^l(+ \ i)^l \frac{\exp[i\varrho]}{i\varrho} = (- \ i)^l \frac{\exp[i\varrho]}{i\varrho}$$
$$= \frac{\exp\{i[\varrho - l(\pi/2)]\}}{i\varrho} .$$

With this we immediately obtain

$$\psi_l = \frac{\sin(\varrho - l\pi/2)}{\varrho} . \qquad [21.10]$$

On the other hand, for $\varrho \ll 1$, all that we need differentiate in [21.8] is $1/\varrho$. The result is

$$\left.\begin{array}{l} \psi_l \sim \dfrac{\varrho^l}{1 \times 3 \times 5 \times \ldots \times (2l+1)} \\ \chi_l \sim -\dfrac{1 \times 3 \times 5 \times \ldots \times (2l-1)}{\varrho^{l+1}} \end{array}\right\} . \qquad [21.11]$$

Connection with the cylinder functions

There is the following relation between the solutions $\xi_l^{(1)(2)}$ and ψ_l and the cylinder functions:

$$\xi_l^{(1)(2)} = \sqrt{\frac{\pi}{2\varrho}}\, H_{l+\frac{1}{2}}^{(1)(2)}(\varrho) , \qquad [21.12]$$

$$\psi_l = \sqrt{\frac{\pi}{2\varrho}}\, I_{l+\frac{1}{2}}(\varrho) . \qquad [21.13]$$

The functions $H_{l+\frac{1}{2}}^{(1)(2)}$ are the Hankel functions of the first and second kinds, and $I_{l+\frac{1}{2}}$ is the Bessel function. It is well known that these functions have the following integral

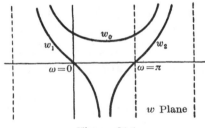

Figure 21.1

representations (see **Fig. 21.1**):

$$I_n(\varrho) = \frac{1}{2\pi}\int_{w_0} \exp\left[i\varrho \cos \omega\right] \exp\left[in\left(\omega - \frac{\pi}{2}\right)\right] \mathrm{d}\omega ,$$

$$H_n^{(1)(2)}(\varrho) = \frac{1}{2\pi}\int_{\substack{w_1 \\ w_2}} \exp\left[i\varrho \cos \omega\right] \exp\left[in\left(\omega - \frac{\pi}{2}\right)\right] \mathrm{d}\omega .$$

Substituting [21.12] or [21.13] into [21.3] does indeed lead to the differential equation for the cylinder functions:

$$\frac{\mathrm{d}^2 Z_n}{\mathrm{d}\varrho^2} + \frac{1}{\varrho}\frac{\mathrm{d}Z_n}{\mathrm{d}\varrho} + \left(1 - \frac{n^2}{\varrho^2}\right)Z_n = 0 . \qquad [21.14]$$

22. EXPANSION OF A PLANE WAVE IN LEGENDRE POLYNOMIALS

We want to expand the plane wave e^{ikz} in terms of Legendre polynomials $(x \equiv \cos\vartheta)$:

$$e^{ikz} = \exp[ikr\cos\vartheta] = e^{i\varrho x} = \sum_l f_l(\varrho)P_l(x). \qquad [22.1]$$

The coefficients $f_l(\varrho)$ are to be determined. In order to do this we need the integral representation

$$S_l = \frac{1}{l!}\left(\frac{\varrho}{2}\right)^l \frac{1}{2}\int_{-1}^{+1} e^{i\varrho x}(1-x^2)^l \, \mathrm{d}x = \psi_l(\varrho), \qquad [22.2]$$

which we want to prove by deriving a recursion formula analogous to [21.4]. We have

$$S_0 = \frac{1}{2}\int_{-1}^{+1} e^{i\varrho x} \, \mathrm{d}x = \frac{e^{i\varrho} - e^{-i\varrho}}{2i\varrho} = \frac{\sin\varrho}{\varrho} = \psi_0(\varrho).$$

We can write S_{l-1} in terms of S_l:

$$S_{l+1} = \frac{1}{(l+1)!}\left(\frac{\varrho}{2}\right)^{l+1}\frac{1}{2}\int_{-1}^{+1} e^{i\varrho x}(1-x^2)^{l+1} \, \mathrm{d}x$$

$$= -\frac{1}{(l+1)!}\left(\frac{\varrho}{2}\right)^{l+1}\frac{1}{2}\int_{-1}^{+1}\frac{e^{i\varrho x}}{i\varrho}(1-x^2)^l(l+1)(-2x) \, \mathrm{d}x$$

$$= -\frac{1}{l!}\left(\frac{\varrho}{2}\right)^l \frac{\mathrm{d}}{\mathrm{d}\varrho}\frac{1}{2}\int_{-1}^{+1} e^{i\varrho x}(1-x^2)^l \, \mathrm{d}x$$

$$= -\varrho^l\frac{\mathrm{d}}{\mathrm{d}\varrho}(\varrho^{-l}S_l) = -\frac{\mathrm{d}}{\mathrm{d}\varrho}S_l + S_l\frac{l}{\varrho}.$$

Thus, since this is the same as the recursion formula [21.4], ψ_l is indeed equal to the integral S_l.

We need the following well-known relations for the Legendre polynomials (see [18.14], [18.16], [18.21]):

$$P_l(x) = \frac{(-1)^l}{2^l l!} \left(\frac{\mathrm{d}}{\mathrm{d}x}\right)^l (1 - x^2)^l,$$

$$\int_{-1}^{+1} P_l(x) P_{l'}(x)\,\mathrm{d}x = \delta_{ll'} \frac{2}{2l+1}, \qquad P_l(1) = 1.$$

Using these relations, we immediately obtain

$$\frac{2}{2l+1} f_l(\varrho) = \int_{-1}^{+1} e^{i\varrho x} P_l(x)\,\mathrm{d}x = \frac{1}{2^l l!} (i\varrho)^l \int_{-1}^{+1} e^{i\varrho x} (1 - x^2)^l\,\mathrm{d}x$$

from [22.1]; the last expression is the result of n-fold partial integration. By comparing with [22.2] we arrive at

$$i^l 2\psi_l(\varrho) = \frac{2}{2l+1} f_l(\varrho),$$

and with this we obtain the desired expansion:

$$\exp[i\varrho \cos\vartheta] = \sum_l (2l+1) i^l \psi_l(\varrho) P_l(\cos\vartheta). \qquad [22.3]$$

23. SOLUTION OF THE SCHRÖDINGER EQUATION WITH AN ARBITRARY CENTRAL POTENTIAL

In the wave equation

$$\nabla^2 u + k^2 u - \frac{2m}{h^2} V(r) u = 0, \qquad [23.1]$$

we no longer use the Coulomb potential, which has special properties; instead, we simply require

$$\lim_{r \to \infty} r V(r) = 0, \qquad [23.2]$$

that is, that the potential $V(r)$ fall off faster than the Coulomb potential. With the usual separation,

$$u = v_l(\varrho) Y_{l,m}(\vartheta, \varphi), \qquad [23.3]$$

and with the abbreviations

$$kr = \varrho, \qquad \frac{2m}{\hbar^2} V(r) = U(r), \qquad [23.4]$$

the equation

$$\frac{1}{\varrho} \frac{d^2}{d\varrho^2} (\varrho v_l) + \left\{ 1 - \frac{l(l+1)}{\varrho^2} \right\} v_l - \frac{U(r)}{k^2} v_l = 0 \qquad [23.5]$$

follows from [23.1].

a. Method of partial waves

We know two special cases of the solution of [23.5]:

$$\varrho \ll 1: \quad v_l(\varrho) \sim \varrho^l \quad \text{if } U(0) \text{ is finite} \qquad [23.6]$$

$\big($this can be verified, for example, by a power series trial solution $\big($see [18.30]$\big)\big)$;

$$\varrho \gg 1: \quad v_l(\varrho) \sim \frac{c}{\varrho} \sin\left(\varrho - l\frac{\pi}{2} + \delta_l(k)\right) \qquad [23.7]$$

(see, for example, [19.12]). The latter, asymptotic representation was introduced in an exercise (see Sec. 46) in which δ_l was defined to be zero for $U = 0$. In the following, the normalization of v_l in this formula will be chosen such that $c = 1$ (this normalization is different from that in Section 46, where c equals $k\sqrt{2/\pi}$). Here we want to set up a different representation which is more practical for our scattering problem.[3] In analogy to formula [19.10], we take

$$u = \exp[i\varrho \cos\vartheta] + f(\vartheta) \frac{\exp[i\varrho]}{r}. \qquad [23.8]$$

This trial solution satisfies two conditions: that the solution not contain an incoming spherical wave, and that it only consist of the incident plane wave at large distances

---

[3] This rather arbitrary normalization is meaningless here, because it only changes the coefficients in [23.9].

from the scattering center. In order to determine $f(\theta)$, we now require that the asymptotic formula [23.8] agree with the general solution of the differential equation [23.1] at large distances, and we write this solution in the form

$$u = \sum_l a_l v_l(\varrho) P_l(\cos \vartheta) \qquad [23.9]$$

(φ does not appear because the problem is axially symmetric). By expanding the plane wave in [23.8] according to [22.3] and, in [23.9], using the asymptotic representation [23.7] for $v_l(\varrho)$, normalized with $c = 1$, we obtain

$$\sum_l (2l+1) i^l \sin\left(\varrho - l\frac{\pi}{2}\right) P_l(\cos \vartheta) + kf(\vartheta) e^{i\varrho}$$

$$= \sum_l a_l \sin\left(\varrho - l\frac{\pi}{2} + \delta_l(k)\right) P_l(\cos \vartheta) \qquad [23.10]$$

by equating the two solutions. We have also used [21.10] here. Now, we write this equation in a different form:

$$\sum_l (2l+1) i^l \frac{1}{2i} \{e^{i\varrho}(-i)^l - e^{-i\varrho} i^l\} P_l(\cos \vartheta) + kf(\vartheta) e^{i\varrho}$$

$$= \sum_l a_l \frac{1}{2i} \{e^{i\varrho} e^{i\delta_l}(-i)^l - e^{-i\varrho} e^{-i\delta_l} i^l\} P_l(\cos \vartheta). \qquad [23.11]$$

Setting the coefficients of $e^{-i\varrho}$ and $e^{i\varrho}$ equal to zero, we obtain

$$a_l = (2l+1) i^l \exp[+i\delta_l], \qquad [23.12]$$

$$f(\vartheta) = \frac{1}{2ik} \sum_l (2l+1)(\exp[2i\delta_l] - 1) P_l(\cos \vartheta)$$

$$= \frac{1}{k} \sum_l (2l+1) \sin \delta_l \exp[i\delta_l] P_l(\cos \vartheta). \qquad [23.13]$$

b. The differential and total scattering cross sections

The *differential* cross section [20.2] can now be written as

$$dQ = |f(\vartheta)|^2 d\Omega$$

$$= \frac{1}{k^2} |\sum_l (2l+1) \sin \delta_l \exp[i\delta_l] P_l(\cos \vartheta)|^2 d\Omega. \qquad [23.14]$$

We obtain the *total* cross section by integrating over solid angle $(d\Omega = \sin\vartheta \, d\vartheta \, d\varphi)$:

$$Q = \int dQ = 2\pi \int_0^\pi |f(\vartheta)|^2 \sin\vartheta \, d\vartheta .$$

All of the cross terms vanish because of the orthogonality of the Legendre polynomials. Thus, using the normalization integral $\int_{-1}^{+1} P_l^2(x) dx = 2/(2l+1)$, we can immediately write

$$Q = \frac{4\pi}{k^2} \sum_l (2l+1) \sin^2 \delta_l(k) . \qquad [23.15]$$

c. Calculation of the phase shift $\delta_l(k)$

We now want to calculate the phase shift $\delta_l(k)$ between the unperturbed wave function, [21.10], and the wave function perturbed by a potential, [23.7]. We multiply the differential equation

$$\frac{1}{r} \frac{d^2}{dr^2} (r\psi_l) + \left(k^2 - \frac{l(l+1)}{r^2} \right) \psi_l = 0 \qquad [23.16]$$

by $-r^2 v_l$ and add to it Eq. [23.5] multiplied by $\varrho^2 \psi_l$:

$$\frac{d}{dr} \left\{ (r\psi_l) \frac{d}{dr} (rv_l) - (rv_l) \frac{d}{dr} (r\psi_l) \right\} = U(r) r^2 \psi_l v_l ,$$

$$\left\{ (r\psi_l) \frac{d}{dr} (rv_l) - (rv_l) \frac{d}{dr} (r\psi_l) \right\}_{r=R} - 0 = \int_0^R U(r) r^2 \psi_l v_l dr. \qquad [23.17]$$

Using the trigonometric identity

$$\sin a \cdot \cos(a+\delta) - \sin(a+\delta) \cos a = -\sin\delta ,$$

we obtain the exact formula

$$\sin\delta_l(k) = -k \int_0^\infty U(r) r^2 \psi_l(kr) v_l(kr) \, dr , \qquad [23.18]$$

in which we unfortunately do not know the v_l. In certain cases we can circumvent this difficulty by substituting ψ_l for v_l as a first approximation:

$$\sin \delta_l(k) = -k \int_0^\infty U(r) r^2 \psi_l^2 \, dr . \qquad [23.19]$$

24. THE BORN APPROXIMATION

Again, we let the solution of Eq. [23.1],

$$\nabla^2 u + k^2 u = U(r) \cdot u , \qquad \text{where} \qquad U(r) = \frac{2m}{h^2} V(r) , \qquad [24.1]$$

have the form

$$u = \exp[ikr \cos \vartheta] + u_1 + \dots , \qquad [24.2]$$

where the perturbation u_1, which represents the scattered wave, is to be small compared with the incident plane wave. Substituting, we obtain

$$\nabla^2 u_1 + k^2 u_1 = U(r) \exp[ikr \cos \vartheta] \qquad [24.3]$$

to lowest order; the term $U(r)u_1$ has been neglected. Of course, we could continue this method and write

$$\nabla^2 u_2 + k^2 u_2 = U(r) \cdot u_1 , \qquad \text{etc.}$$

However, it turns out that the Born approximation is practical only if the lowest approximation is sufficient.

The solution of Eq. [24.3] is worked out in the theory of the Hertz dipole[4] (Fig. 24.1); it is

$$u_1(x) = -\frac{1}{4\pi} \int \frac{U(r') \exp[ikr' \cos \vartheta']}{r''} \exp[ikr''] \, dV' . \qquad [24.4]$$

It contains the radiation boundary condition which is also

---

[4] See W. PAULI, *Lectures in Physics: Electrodynamics* (M.I.T. Press, Cambridge, Mass., 1972).

needed here. We do not need this solution in all general-
ity; we restrict ourselves to two special cases.

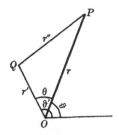

Figure 24.1

P: field point; Q: source point; O: origin

We first investigate what happens when $P \to 0$:

$$u_1(0) = -\frac{1}{4\pi}\int \frac{U(r')\exp[ikr'(\cos\vartheta' + 1)]}{r'}\,dV'. \qquad [24.5]$$

If we integrate over solid angle $(dV' = r'^2 \sin\vartheta'\,dr'\,d\vartheta'\,d\varphi')$
and omit the primes, we obtain

$$u_1(0) = -\frac{1}{k}\int U(r)\sin kr \exp[ikr]\,dr. \qquad [24.6]$$

This expression represents the scattered wave at the origin.
By assumption, its amplitude is much less than 1 (we have
normalized the incident wave to 1), which will certainly
be true if

$$\frac{1}{k}\int |U(r)|\,dr \ll 1. \qquad [24.7]$$

This is a sufficient condition for the applicability of perturbation theory.[5]

Now, let us investigate the behavior of the solution [24.4]

---

[5] Therefore, the Born approximation is to be used if the kinetic energy
of the incident particle is large compared to the interaction energy. It
thus complements the method of partial waves, which is valid at low energies.

at large distances:

$$r \gg d = \text{range of the potential} . \qquad [24.8]$$

With

$$r'' = r - r' \cos \theta , \qquad [24.9]$$

we obtain

$$u_1 = - \frac{\exp[ikr]}{r} \frac{1}{4\pi} \int U(r') \exp\left[ikr'(\cos\vartheta' - \cos\theta)\right] \mathrm{d}V' ,$$

or, by expanding the plane waves in Legendre polynomials according to [22.3],

$$u_1 = - \frac{\exp[ikr]}{r} \frac{1}{4\pi} \int U(r')r'^2 \mathrm{d}r' \sum_{l,l'} i^l(2l+1)\psi_l(kr')(-i)^{l'}$$

$$\times (2l'+1)\psi_{l'}(kr') \int P_l(\cos\vartheta') P_{l'}(\cos\theta) \, \mathrm{d}\Omega' . \qquad [24.10]$$

Because of the orthogonality of the Legendre polynomials (also for the angles ϑ', θ of Fig. 24.1), only the term with $l' = l$ is nonzero:

$$u_1 = - \frac{\exp[ikr]}{r} \int U(r')r'^2 \, \mathrm{d}r' \sum_l (2l+1)^2$$

$$\times \{\psi_l(kr')\}^2 \int \frac{\mathrm{d}\Omega'}{4\pi} P_l(\cos\vartheta') P_l(\cos\theta) . \qquad [24.11]$$

Using the relation

$$P_l(\cos\vartheta) = (2l+1) \int P_l(\cos\theta) P_l(\cos\vartheta') \frac{\mathrm{d}\Omega'}{4\pi} \qquad [24.12]$$

which holds for the angles of Fig. 24.1 we then have

$$u_1 = - \frac{\exp[ikr]}{r} \sum_l (2l+1) P_l(\cos\vartheta)$$

$$\times \int_0^\infty U(r')r'^2 \{\psi_l(kr')\}^2 \, \mathrm{d}r' . \qquad [24.13]$$

This result matches the one which we already had, except that a factor $e^{i\delta_l}$ has been neglected here, because δ_l is very small (see [23.8], [23.13], and [23.19]).

25. SCATTERING OF LOW-ENERGY PARTICLES

To conclude our treatment of collision theory, we briefly consider a calculational method that uses the following two assumptions: (1) $U(r) \sim 0$ for $r \gg a$ (a need not be precisely defined); (2) $ka \ll 1$, that is, the energy of the incident particle is small. For $r \gg a$, the term $U(r)v_l$ in the differential equation

$$\frac{1}{r}\frac{d^2}{dr^2}(rv_l) + \left(k^2 - \frac{l(l+1)}{r^2}\right)v_l = U(r)v_l \qquad [25.1]$$

can be neglected, and we have the two independent solutions of the force-free equation ([21.5]-[21.9]),

$$\psi_l = \frac{\xi^{(1)} + \xi^{(2)}}{2} \qquad [25.2]$$

and

$$\chi_l = \frac{\xi^{(1)} - \xi^{(2)}}{2i}. \qquad [25.3]$$

The asymptotic forms of these solutions for $kr \gg 1$ are

$$\psi_l \sim \frac{1}{kr}\sin\left(kr - l\frac{\pi}{2}\right), \qquad [25.4]$$

$$\chi_l \sim -\frac{1}{kr}\cos\left(kr - l\frac{\pi}{2}\right). \qquad [25.5]$$

The general solution of [25.1] has the form

$$v_l = A\psi_l(kr) + B\chi_l(kr). \qquad [25.6]$$

Comparison with the asymptotic formula [23.7],

$$v_l \sim \frac{1}{kr}\sin\left(kr - l\frac{\pi}{2} + \delta_l\right), \qquad [25.7]$$

yields

$$A = \cos\delta_l, \qquad B = -\sin\delta_l. \qquad [25.8]$$

In contrast with what we did before $(r \gg a, kr \gg 1)$, we now assume $ka \ll kr \ll 1$. Because of $k \ll 1/r$, [25.1] reduces further to

$$\frac{d^2}{dr^2}(rv_l) - \frac{l(l+1)}{r^2}(rv_l) = 0. \qquad [25.9]$$

The differential equation has the solution

$$v_l = c_1 \left(\frac{r}{a}\right)^l + c_2 \left(\frac{r}{a}\right)^{-l-1} \qquad (c_1/c_2 \text{ independent of } k). \qquad [25.10]$$

On the other hand, the general solution [25.6] with the (constant!) coefficients [25.8] is still valid; we just have to substitute the approximations [21.11], as a result of $kr \ll 1$. Then we obtain

$$v_l = \cos\delta_l \cdot \psi_l(kr) - \sin\delta_l \cdot \chi_l(kr) = \cos\delta_l \frac{(kr)^l}{1 \times 3 \times \ldots \times (2l+1)}$$
$$+ 1 \times 3 \times \ldots \times (2l-1)\,\sin\delta_l \cdot (kr)^{-l-1}. \qquad [25.11]$$

Comparison of the two solutions yields

$$1 \times 3 \times \ldots \times (2l-1) \times 1 \times 3 \times \ldots \times (2l+1)(ka)^{-2l-1}\tan\delta_l$$

$$= \frac{c_2}{c_1} = \lambda \text{ (free parameter, independent of } k),$$

or

$$\tan\delta_l = \lambda \frac{(ka)^{2l+1}}{\{1 \times 3 \times \ldots \times (2l-1)\}^2(2l+1)}; \qquad [25.12]$$

since δ_l is small, we can write

$$\frac{\sin\delta_l}{k} \sim a \cdot \lambda \frac{(ka)^{2l}}{\{1 \times 3 \times \ldots \times (2l-1)\}^2(2l+1)}. \qquad [25.13]$$

This relation shows that the amplitude of the lth partial

wave in [23.13] is of the order of magnitude of $(ka)^{2l}$. If this expression is substituted in the formula for the total cross section [23.15],

$$Q = 4\pi \sum_l (2l+1) \frac{\sin^2 \delta_l}{k^2},$$

then we see, for example, that the isotropic part of the scattering ($l=0$, s wave) is independent of the energy of the incident particle at low energies.

Chapter 7. Approximate Methods for Solving the Wave Equation

26. EIGENVALUE PROBLEM OF A PARTICLE IN A UNIFORM FIELD

a. The Airy function

We imagine a particle in a uniform field:

$$E_{\text{pot}} = - F \cdot q \,. \qquad [26.1]$$

The one-dimensional wave equation,

$$\frac{\mathrm{d}^2 u}{\mathrm{d}q^2} + \frac{2m}{h^2} \left(Fq + E \right) u = 0 \,, \qquad [26.2]$$

can, in terms of new variables

$$x = \sqrt[3]{\frac{2mF}{h^2}} \left(q + q_0 \right), \qquad E = Fq_0 \,, \qquad [26.3]$$

be written in the simpler form

$$\frac{\mathrm{d}^2 u}{\mathrm{d}x^2} + xu = 0 \,. \qquad [26.4]$$

We solve this equation, which is familiar from the theory of diffraction, with the trial solution

$$u = \int_C e^{xt} f(t) \, \mathrm{d}t \,. \qquad [26.5]$$

Using

$$u'' = \int_C e^{xt} t^2 f \, \mathrm{d}t \,,$$

$$xu = \int_C \left(\frac{\mathrm{d}}{\mathrm{d}t} e^{xt} \right) f \, \mathrm{d}t = \int_C \frac{\mathrm{d}}{\mathrm{d}t} \left(e^{xt} f \right) \mathrm{d}t - \int_C e^{xt} f' \, \mathrm{d}t$$

126

and [26.4], we obtain

$$t^2 f - f' = 0, \qquad \frac{f'}{f} = t^2, \qquad f = \text{constant} \times \exp\left[\frac{t^3}{3}\right], \qquad [26.6]$$

as long as the path of integration C is chosen such that

$$\int_C \frac{\mathrm{d}}{\mathrm{d}t}\left(e^{xt} f\right) \mathrm{d}t = 0. \qquad [26.7]$$

Now, the solution can be written, with the usual normalization, as

$$u_k = -\frac{i}{\pi} \int_{C_k} \exp\left[xt + \frac{t^3}{3}\right] \mathrm{d}t. \qquad [26.8]$$

Condition [26.7] is satisfied and convergence of the solution is guaranteed if, for $|t| \to \infty$, we require

$$\cos 3\varphi < 0 \qquad \left(t = |t|\, e^{i\varphi}\right),$$

that is

$$2\pi n + \frac{\pi}{2} \leqslant 3\varphi \leqslant \frac{3\pi}{2} + 2\pi n.$$

For example, the good intervals are

$$\left(\frac{\pi}{6}, \frac{\pi}{2}\right), \quad \left(\frac{5\pi}{6}, \frac{7\pi}{6}\right), \quad \left(\frac{9\pi}{6}, \frac{11\pi}{6}\right).$$

The regions corresponding to these intervals are shaded

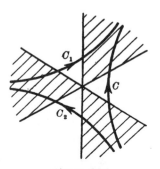

Figure 26.1

in Fig. 26.1. It is evident that

$$\int_C = \int_{C_1} + \int_{C_2} ,$$

that is, there exist two independent particular integrals. The solutions u_k are called the Airy functions and they are denoted by $A^{(k)}$. They are well known to us from the theory of diffraction.[1] If x is real, then $A^{(1)}$ and $A^{(2)}$ are complex conjugates. Later we shall also need the average,

$$A = \tfrac{1}{2}(A^{(1)} + A^{(2)}) . \tag{26.9}$$

Finally, let it be noted that the Airy functions are related to the cylinder functions:

$$A^{(k)} = C^{(k)}\sqrt{x} \cdot H_{\frac{1}{3}}^{(k)}\left(\frac{2}{3}x^{\frac{3}{2}}\right), \qquad C^{(1)} = \frac{\exp[i\pi/6]}{\sqrt{3}} = C^{(2)*}.$$

By means of the saddle point method, we shall now derive asymptotic expressions from the integral [26.8].

b. The saddle point method

In order to evaluate the integral

$$\int_C \exp[f(t)]\,dt , \tag{26.10}$$

we first expand $f(t)$ about the "saddle point" t_0 which is defined by

$$f'(t_0) = 0:$$

the expression that we use is

$$f(t) = f(t_0) + \frac{(t-t_0)^2}{2}f''(t_0) + \dots . \tag{26.11}$$

In order that $\mathrm{Re}(f)$ fall off as rapidly as possible, we choose

---

[1] See W. Pauli, *Lectures in Physics: Optics and the Theory of Electrons* (M.I.T. Press, Cambridge, Mass., 1972).

the path of integration to pass through t_0 (following the method of steepest descent of Cauchy and Riemann); then we have $\mathrm{Im}(f) = \mathrm{constant}$. Let

$$f''(t_0) = |f''(t_0)| e^{i\alpha} . \qquad [26.12]$$

Because the imaginary part is constant, we then have

$$(t - t_0)^2 = -\varrho^2 e^{i\alpha}$$

$$t - t_0 = \varrho \exp\left[i\frac{\pi - \alpha}{2}\right]$$

$$dt = d\varrho \exp\left[i\frac{\pi - \alpha}{2}\right],$$

and, as a result,

$$f(t) = f(t_0) - \tfrac{1}{2}|f''(t_0)|\varrho^2 . \qquad [26.13]$$

Let us be bold and omit the higher terms of the expansion [26.11] when we evaluate the integral [26.10]. In this approximation we then obtain

$$\int_\sigma \exp[f(t)]dt = \exp[f(t_0)]\exp\left[i\frac{\pi - \alpha}{2}\right]\int \exp[-\tfrac{1}{2}|f''(t_0)|\varrho^2]d\varrho$$

$$= \exp[f(t_0)]\exp\left[i\frac{\pi - \alpha}{2}\right]\sqrt{\frac{2\pi}{|f''(t_0)|}},$$

and, finally,

$$\int_\sigma \exp[f(t)]\,dt = \exp[f(t_0)]\sqrt{\frac{2\pi}{-f''(t_0)}} \qquad [26.14]$$

with t_0 defined by $f'(t_0) = 0$. According to [26.12], an appropriate sign still must be chosen for the square root. For an exact justification of the above calculations see, for example, Courant and Hilbert, *Methods of Mathematical Physics*.

c. Approximate solution for a particle in a homogeneous field

First, we must determine t_0:

$$f(t) = xt + \frac{t^3}{3}, \quad f'(t_0) = x + t_0^2 = 0, \quad f''(t_0) = 2t_0, \quad t_0^2 = -x;$$

$$t_0 = \pm i\sqrt{x}, \qquad \text{for} \quad x > 0, \qquad [26.15]$$

$$t_0 = \pm \sqrt{|x|}, \qquad \text{for} \quad x < 0. \qquad [26.16]$$

For the case $x < 0$, because of

$$f(t_0) = -\tfrac{2}{3} t_0^3 = \pm \tfrac{2}{3} |x|^{\frac{3}{2}},$$

there are two terms in the solution [26.8]:

$$A^{(1)} = -\frac{i}{\sqrt{\pi}} |x|^{-\frac{1}{4}} \exp\left[+\tfrac{2}{3}|x|^{\frac{3}{2}}\right]$$

$$+ \frac{1}{2} \frac{1}{\sqrt{\pi}} |x|^{-\frac{1}{4}} \exp\left[-\tfrac{2}{3}|x|^{\frac{3}{2}}\right], \qquad [26.17]$$

$$A^{(2)} = +\frac{i}{\sqrt{\pi}} |x|^{-\frac{1}{4}} \exp\left[+\tfrac{2}{3}|x|^{\frac{3}{2}}\right]$$

$$+ \frac{1}{2} \frac{1}{\sqrt{\pi}} |x|^{-\frac{1}{4}} \exp\left[-\tfrac{2}{3}|x|^{\frac{3}{2}}\right]. \qquad [26.18]$$

This is because the path of integration goes over two saddle points, as shown in Fig. 26.2. Of course, the right-hand saddle is only half covered and, for that reason, its con-

Path of integration: $\text{Im}(f) = $ constant.

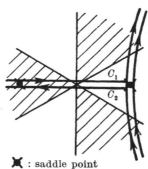

$\rlap{\times}{\blacktriangleright}$: saddle point

Figure 26.2

tribution to the solutions is only half of what it would have been otherwise.

In the case $x > 0$, we have only one saddle point on each of the paths of integration C_1 and C_2 (see Fig. 26.3). The

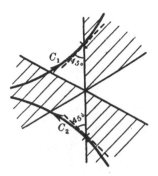

Figure 26.3

solutions (with the famous $\exp[i(\pi/4)]$) are:

$$A^{(1)} = \frac{1}{\sqrt{\pi}} x^{-\frac{1}{4}} \exp[i(\tfrac{2}{3} x^{\frac{3}{2}} - \tfrac{1}{4}\pi)], \qquad [26.19]$$

$$A^{(2)} = \frac{1}{\sqrt{\pi}} x^{-\frac{1}{4}} \exp[-i(\tfrac{2}{3} x^{\frac{3}{2}} - \tfrac{1}{4}\pi)]. \qquad [26.20]$$

Therefore, the average values [26.9] of the two functions are

$$x < 0: \qquad A = \frac{1}{2}\frac{1}{\sqrt{\pi}}|x|^{-\frac{1}{4}} \exp[-\tfrac{2}{3}|x|^{\frac{3}{2}}], \qquad [26.21]$$

$$x > 0: \qquad A = \frac{1}{\sqrt{\pi}} x^{-\frac{1}{4}} \cos(\tfrac{2}{3} x^{\frac{3}{2}} - \tfrac{1}{4}\pi). \qquad [26.22]$$

It must be emphasized again that this solution represents an approximation. The Taylor expansion [26.11] is only appropriate in the neighborhood of the saddle, although the integration [26.8] runs from minus infinity to plus infinity. Of course, it is true that the major contribution to the integral comes from the region of the saddle. We further note that the first terms of $A^{(1)}$ and $A^{(2)}$ in [26.17] and [26.18] are critical, because the error is of the order of magnitude of the expression itself. However, these terms cancel in [26.21].

27. THE WKB METHOD

G. Wentzel and L. Brillouin have given a procedure with which it is possible to obtain the approximate solution of the wave equation when it is separable in the individual independent variables.[2] H. A. Kramers and his students gave the mathematical foundation of the method and extended it.[3] The formalism for obtaining the solution is analogous to the method in optics which leads from the general wave equation to the eikonal equation.[4]

We solve the one-dimensional and time-independent wave equation,

$$u'' + \frac{2m}{h^2}[E - V(x)]u = 0 , \quad \text{where} \quad E - V(x) > 0 , \quad [27.1]$$

with the trial solution

$$u = \exp\left[\frac{i}{h} S\right] . \quad [27.2]$$

In this way, from the second-order differential equation of

---

[2] L. BRILLOUIN, *Compt. Rend.* **183**, 24 (1926); *J. de Physique* **7**, 353 (1926); G. WENTZEL, *Z. Physik* **38**, 518 (1926); also see H. JEFFREYS, *Proc. London Math. Soc.* (2) **23**, 428 (1923).

[3] H. A. KRAMERS, *Z. Physik* **39**, 828 (1926).

[4] In optics (see Ref. 1 on p. 127) we solve the time-independent wave equation,

$$\frac{\partial^2 \psi}{\partial x^2} + n^2 k_0^2 \psi = 0 ,$$

with the trial solution

$$\psi = \exp[ik_0 S] .$$

We assume S to be a slowly varying quantity; namely, only the largest power of k_0 in the expression

$$\frac{\partial^2 \psi}{\partial x^2} = \left\{- k_0^2 \left(\frac{\partial S}{\partial x}\right)^2 + ik_0 \frac{\partial^2 S}{\partial x^2}\right\} \psi$$

is used. From this follows the differential equation for the eikonal which is characteristic of ray optics,

$$\left(\frac{\partial S}{\partial x}\right)^2 = n^2 .$$

the first degree for u, Eq. [27.1], we obtain one of first order and second degree for $dS/dx = S'$ (Riccati differential equation):

$$S'^2 = 2m[E - V(x)] + ihS''. \tag{27.3}$$

To solve this equation, we expand S in powers of h/i (this power series is of the asymptotic type):

$$S = S_0 + \frac{h}{i} S_1 + \dots . \tag{27.4}$$

Substituting [27.4] into [27.3] then yields

$$S_0' = \sqrt{2m[E - V(x)]}, \tag{27.5}$$

which leads to the solution

$$S_0 = \int_{x_1}^{x} \sqrt{2m[E - V(x)]}\, dx . \tag{27.6}$$

If we set

$$p = S_0',$$

then, to zeroth order ($h = 0$), we obtain the relation

$$p(x) = \sqrt{2m[E - V(x)]}, \tag{27.7}$$

which is known from classical mechanics.

The trajectory of the particle is classical in the interval (x_1, x_2) in which

$$E - V(x) \geqslant 0 .$$

It is also possible for there to be many such intervals; however, for the present we assume only one (Fig. 27.1).

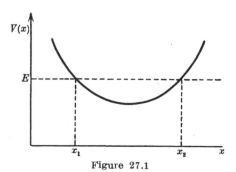

Figure 27.1

From [27.3], the next approximation is

$$2S_1'S_0' + S_0'' = 0 \, ,$$

$$S_1 = -\tfrac{1}{2} \log S_0' + \text{constant} \, .$$

For the solution to the wave equation [27.1] we now obtain

$$u_\pm = \frac{\text{constant}}{\sqrt{p(x)}} \times \exp\left[\frac{i}{h} \cdot \int_{x_1}^{x} p(x) \, \mathrm{d}x\right]; \qquad [27.8]$$

$E - V > 0$: oscillatory (classically attainable),

$E - V < 0$: damped (classically not attainable).

The $\pm$ is to remind us that we are concerned with two linearly independent solutions, corresponding to the two signs of the square root [27.7]. This solution is not usable in the vicinity of

$$p(x) \equiv \sqrt{2m[E - V(x)]} = 0 \; ; \qquad [27.9]$$

that is, it is not valid in the neighborhood of the classical turning point, where, indeed, $E = V(x)$.

In this region we can make a connection with the Airy functions that we have already seen. For this, the expansion

$$\frac{p^2(x)}{2m} = E - V(x) = (x - x_1)F_1 + \dots, \quad F_1 \equiv -V'(x_1) > 0, \; [27.10]$$

must be substituted into Eq. [27.1]:

$$u'' + \frac{2mF_1}{h^2}(x - x_1)u = 0 \, . \qquad [27.11]$$

The solutions of this equation are Airy functions of the variable

$$\xi = (x - x_1) \cdot \sqrt[3]{\frac{2mF_1}{h^2}} \, . \qquad [27.12]$$

In order to find the correct combination of $A^{(1)}$ and $A^{(2)}$, we first consider the region $x < x_1$, in which u must be damped (classically not attainable). With $p = -i|p|$, [27.8] reads

$$u = \frac{\text{constant}}{\sqrt{|p|}} \times \exp\left[-\frac{1}{h}\int_x^{x_1}|p|\,dx\right], \qquad x < x_1. \qquad [27.13]$$

From [27.10] and [27.12] we obtain

$$\frac{1}{h}\int_x^{x_1}|p|\,dx = \sqrt{\frac{2mF_1}{h^2}}\int_x^{x_1}|x-x_1|^{\frac{1}{2}}\,dx = \tfrac{2}{3}|\xi|^{\frac{3}{2}}.$$

Then, according to [26.21], we have, for $x < x_1$,

$$u = \text{constant} \times |\xi|^{-\frac{1}{4}}\exp\left[-\tfrac{2}{3}|\xi|^{\frac{3}{2}}\right] = \text{constant} \times A(\xi),$$

and the solution for $x > x_1$ is given by [26.22] as

$$u = \text{constant} \times \xi^{-\frac{1}{4}}\cos\left(\tfrac{2}{3}\xi^{\frac{3}{2}} - \tfrac{1}{4}\pi\right)$$

$$= \frac{\text{constant}}{\sqrt{p}} \times \cos\left(\frac{1}{h}\int_{x_1}^{x}p\,dx - \tfrac{1}{4}\pi\right). \qquad [27.14]$$

Similar formulas are valid for the second turning point, x_2. With

$$\frac{p^2(x)}{2m} = E - V(x) = -(x-x_2)F_2 + \dots,$$

$$F_2 \equiv +V'(x_2) > 0, \qquad \eta = \sqrt[3]{\frac{2mF_2}{h^2}}(x_2 - x),$$

we obtain

$$u = \frac{\text{constant}}{\sqrt{|p|}} \times \exp\left[-\frac{1}{h}\int_{x_2}^{x}|p|\,dx\right] = \text{constant}$$

$$\times |\eta|^{-\frac{1}{4}}\exp\left[-\tfrac{2}{3}|\eta|^{\frac{3}{2}}\right] = \text{constant} \times A(\eta) \qquad [27.15]$$

in the region $x > x_2$, and we then have

$$u = \text{constant} \times \eta^{-\frac{1}{4}} \cos \left(\tfrac{2}{3} \eta^{\frac{3}{2}} - \tfrac{1}{4} \pi \right)$$

$$= \frac{\text{constant}}{\sqrt{p}} \times \cos \left(\frac{1}{h} \int_x^{x_2} p \, dx - \tfrac{1}{4} \pi \right) \qquad [27.16]$$

for $x < x_2$. These expressions represent the wave function in the region from $-\infty$ to $+\infty$. The wave function vanishes exponentially at these limits. It changes its structure basically at the two classical turning points, x_1 and x_2. For a more detailed discussion of the behavior of these solutions, reference should be made to the original papers by Kramers and his co-workers. In order to obtain a unique description with the two pairs of solutions—[27.13], [27.14] and [27.15], [27.16]—we must also require

$$\cos \left(\frac{1}{h} \int_{x_1}^x p(x) \, dx - \tfrac{1}{4} \pi \right) = \pm \cos \left(\frac{1}{h} \int_x^{x_2} p(x) \, dx - \tfrac{1}{4} \pi \right), \qquad [27.17]$$

$$\frac{1}{h} \int_{x_1}^{x_2} p(x) \, dx - \frac{\pi}{2} = n\pi \qquad [27.18]$$

in the common interval (x_1, x_2). If we denote the famous phase integral by

$$J = \oint p_x \, dx = 2 \int_{x_1}^{x_2} \sqrt{2m(E - V(x))} \, dx \, ,$$

we no longer obtain the old Bohr quantum condition,

$$J = n \cdot 2\pi h \, ; \qquad [27.19]$$

rather, we obtain the refined condition

$$J = (n + \tfrac{1}{2}) \cdot 2\pi h \qquad (n = 0, 1, 2, \ldots) \, . \qquad [27.20]$$

Of course, because of our derivation, even this expression

is an approximation, which becomes better as n increases. The error is of order $1/n$; however, the constants are correct. In some cases, as, for example, with the harmonic oscillator (see [15.20]), the quantum condition [27.20] is obeyed exactly for arbitrary n; this depends on the particular form of the potential.[5]

Until now, we always assumed only one interval in which $E - V(x) > 0$; this was an essential assumption for our derivation. If there exist more than one such interval, then new effects appear because of the fact that the wave function does not vanish in the classically unattainable intermediate regions (see [27.13] and [27.15]). Therefore, it is possible for the wave function to "leak" from a region (x_1, x_2) to another region (x_3, x_4) in spite of their being separated by a classically insurmountable potential

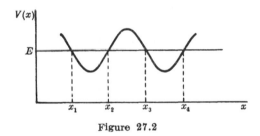

Figure 27.2

barrier. This typically wave-mechanical phenomenon is called the *tunnel effect*; it plays a large role in numerous applications of quantum theory. We shall concern ourselves with the effect in an exercise (see Sec. 41).

[5] See, for example, E. C. Kemble, *The Fundamental Principles of Quantum Mechanics*, pp. 574 ff.

Chapter 8. Matrices and Operators. Perturbation Theory

We recall the discussion of Section 16 and repeat the definition of a matrix element given there. Let

$$u_1(q), \ldots, u_n(q), \ldots$$

be a complete, orthonormal set of functions. Then we can write

$$\underline{F}u_n = \sum_k u_k(k\,|\,\underline{F}\,|\,n) , \qquad [28.1]$$

where

$$(k\,|\,\underline{F}\,|\,n) = \int u_k^*(\underline{F}u_n)\,\mathrm{d}q \qquad [28.2]$$

represents the matrix element of the operator $\underline{F}$ with respect to the u_n.

According to [28.2], the Hermiticity condition [12.5],

$$\int (\underline{F}u_k)^* u_n\,\mathrm{d}q = \int u_k^*(\underline{F}u_n)\,\mathrm{d}q \qquad [28.3]$$

means

$$(k\,|\,\underline{F}\,|\,n) = (n\,|\,\underline{F}\,|\,k)^* \qquad [28.4]$$

for the matrices. Defining another matrix analogously,

$$\underline{G}u_k = \sum_m u_m(m\,|\,\underline{G}\,|\,k) , \qquad (k\,|\,\underline{G}\,|\,m) = \int u_k^*(\underline{G}u_m)\,\mathrm{d}q , \qquad [28.5]$$

we calculate the matrix of the operator $\underline{G}\,\underline{F}$:

$$(\underline{G}\underline{F})u = \underline{G}(\underline{F}u) , \qquad [28.6]$$

$$\underline{G}(\underline{F}u_n) = \sum_k (\underline{G}u_k)(k\,|\,\underline{F}\,|\,n)$$

$$= \sum_m u_m \sum_k (m\,|\,\underline{G}\,|\,k)(k\,|\,\underline{F}\,|\,n) \equiv \sum_m u_m(m\,|\,\underline{G}\underline{F}\,|\,n) . \qquad [28.7]$$

The last step in [28.7] was carried out in the usual sense of matrix multiplication,

$$(m\,|\,\underline{G}\underline{F}\,|\,n) \equiv \sum_k (m\,|\,\underline{G}\,|\,k)(k\,|\,\underline{F}\,|\,n) , \qquad [28.8]$$

which shows that multiplication of operators corresponds to multiplication of the associated matrices.

We now want to investigate how the matrices change if we go from the u_n to another complete, orthogonal set of functions v_A. We can expand v_A in terms of the u_n:

$$v_A = \sum_n u_n(n\,|\,\underline{S}\,|\,A) \equiv \underline{S}u_A , \qquad [28.9]$$

where

$$(n\,|\,\underline{S}\,|\,A) = \int v_A u_n^* \, dq . \qquad [28.10]$$

With each function $f = a_1u_1 + a_2u_2 + ...$, the operator $\underline{S}$ defined in [28.9] associates a function $g = a_1v_1 + a_2v_2 + ...$ with the same expansion coefficients. We call $\underline{S}$ a *transformation operator*; it is fundamentally different from the Hermitian operators $\underline{F}$ considered so far.

a. Unitary transformations

As a result of the completeness relation [9.9], we have

$$\int v_A^* v_B \, dq = \sum_n (n\,|\,\underline{S}\,|\,A)^* (n\,|\,\underline{S}\,|\,B) . \qquad [28.11]$$

With this relation, the orthogonality and normalization

condition for the v_A becomes

$$\sum_n (n \mid \underline{S} \mid A)^* (n \mid \underline{S} \mid B) = (A \mid \underline{S}^\dagger \underline{S} \mid B) = \delta_{AB}$$

or
$$\underline{S}^\dagger \underline{S} = 1$$
$$[28.12]$$

where $\underline{S}^\dagger$ denotes the Hermitian conjugate matrix obtained from $\underline{S}$:

$$(A \mid \underline{S}^\dagger \mid n) \equiv (n \mid \underline{S} \mid A)^* . \qquad [28.13]$$

A rule for the Hermitian conjugate matrix is

$$(\underline{F}\,\underline{G}\,\underline{H}\ldots)^\dagger = \ldots \underline{H}^\dagger \underline{G}^\dagger \underline{F}^\dagger . \qquad [28.14]$$

We can require, as a condition for the completeness of the v_A, that each u_n be expandable in terms of the v_A:

$$u_n = \sum_A v_A (A \mid \underline{S}^\dagger \mid n) , \qquad (A \mid \underline{S}^\dagger \mid n) = \int u_n v_A^* \, dq . \qquad [28.15]$$

If the v_A are to be complete, then the condition analogous to [28.12] must hold:

$$(n \mid \underline{S}\underline{S}^\dagger \mid m) = \delta_{nm} , \qquad \underline{S}\underline{S}^\dagger = 1 ; \qquad [28.16]$$

this means that $\underline{S}^{-1}$, the operator inverse to $\underline{S}$, exists. For square matrices of finite rank this is always the case; that is, [28.16] always follows from [28.12] and vice versa. However, this is not the case with rectangular matrices and, therefore, it is also not the case with matrices of infinite rank, since then it no longer makes sense to speak of a square matrix.

A matrix which satisfies [28.12] and [28.16] is called *unitary*, and the associated transformation is called a *unitary transformation*. A unitary transformation preserves orthogonality, normalization, and completeness.

We now want to investigate how the matrix $\underline{F}$ defined in [28.2] transforms when we go from the u_n to the v_A. Using [28.9], and in analogy with

$$(k \mid \underline{F} \mid l) = \int u_k^* (\underline{F} u_l) \, dq , \qquad [28.17]$$

we write

$$(A \mid \underline{F'} \mid B) = \int v_A^*(\underline{F} v_B)\, dq$$

$$= \int \sum_n \sum_m (A \mid \underline{S}^\dagger \mid n) u_n^*(\underline{F} u_m)(m \mid \underline{S} \mid B)\, dq$$

$$= \sum_n \sum_m (A \mid \underline{S}^\dagger \mid n)(n \mid \underline{F} \mid m)(m \mid \underline{S} \mid B) = (A \mid \underline{S}^\dagger \underline{F} \underline{S} \mid B). \quad [28.18]$$

Thus,

$$\underline{F'} = \underline{S}^\dagger \underline{F} \underline{S} \quad \text{(similarity transformation)} \quad [28.19]$$

and, since $\underline{S}^\dagger = \underline{S}^{-1}$ for a unitary transformation,

$$\underline{F'} = \underline{S}^{-1} \underline{F} \underline{S}. \quad [28.20]$$

It is easily shown that the Hermiticity of $\underline{F}$ is preserved by this transformation; that is, $\underline{F'}$ is Hermitian if $\underline{F}$ is.

b. *Formulation of the eigenvalue problem*

If we write a matrix element of the Hamiltonian operator

$$(n \mid \underline{H} \mid m) = \int u_n^*(H u_m)\, dq \quad [28.21]$$

with respect to eigenfunctions of the wave equation,

$$\underline{H} u = E u,$$

we obtain

$$(n \mid \underline{H} \mid m) = E_m \int u_n^* u_m\, dq = E_m \delta_{nm} ; \quad [28.22]$$

that is, the Hamiltonian matrix defined in this system is diagonal and its diagonal elements are the energy eigenvalues E_n of the wave equation. Thus, solving the wave equation corresponds to diagonalizing the Hamiltonian matrix (written with respect to some orthonormal set of functions); that is, we make a principal axis transformation with a unitary transformation matrix $\underline{S}$ which must satisfy

$$(n \mid \underline{S}^\dagger \underline{H} \underline{S} \mid m) = E_n \delta_{nm}, \quad [28.23]$$

according to [28.19]. Using $\underline{S}\,\underline{S}^\dagger = 1$, we can also write

$$(n\,|\,\underline{H}\underline{S}\,|\,m) = (n\,|\,\underline{S}\underline{E}\,|\,m)\,, \qquad [28.24]$$

or

$$\sum_k (n\,|\,\underline{H}\,|\,k)(k\,|\,\underline{S}\,|\,m) = (n\,|\,\underline{S}\,|\,m)E_m\,. \qquad [28.25]$$

In this fashion it is, in principle, possible not only to calculate the energy eigenvalues but also to calculate the associated eigenfunctions u_n of the wave equation; nevertheless, the calculation can be carried out in only a few cases, for example, with the harmonic oscillator and in perturbation theory.

c. Extension of the matrix method to continuous spectra

The operations of matrix calculus can be generalized to the case in which continuous variables replace the discrete indices. If, for example, we consider the matrix product $\underline{F}\cdot\underline{G}$,

$$\sum_k (n\,|\,\underline{F}\,|\,k)(k\,|\,\underline{G}\,|\,m)\,, \qquad [28.26]$$

and assume k to be a continuous variable, then we need only replace the sum by an integral:

$$\sum_k \to \int \varrho(k_1,\,\ldots,\,k_f)\,\mathrm{d}^f k\,. \qquad [28.27]$$

In place of the Kronecker δ, we have the Dirac function:

$$\delta_{kk'} \to \varrho^{-1}(k)\delta^f(k-k')\,. \qquad [28.28]$$

In this way we can, for example, rewrite conditions [28.12] and [28.16]:

$$\underline{S}^\dagger\underline{S} = 1\,, \quad \sum_n (A\,|\,\underline{S}^\dagger\,|\,n)(n\,|\,\underline{S}\,|\,B) = \varrho^{-1}(A)\delta(A-B)\,; \qquad [28.29]$$

$$\underline{S}\,\underline{S}^\dagger = 1\,, \quad \int (n\,|\,\underline{S}\,|\,A)\varrho(A)\,\mathrm{d}A(A\,|\,\underline{S}^\dagger\,|\,m) = \delta_{nm}\,. \qquad [28.30]$$

Since the "density function" ϱ can also have a discrete spectrum, it is even possible to have a juxtaposition of discrete and continuous spectra.

29. GENERAL FORMALISM OF PERTURBATION THEORY IN THE MATRIX REPRESENTATION

We recall the example in Section 16 of a linear oscillator with a perturbing potential, and we make a more general assertion: We can carry out a perturbation calculation if the Hamiltonian matrix has the form

$$\underline{H} = \underline{H}^{(0)} + \underline{V}, \qquad [29.1]$$

where $\underline{H}^{(0)}$ is diagonal, and where $\underline{V}$ is not diagonal but is small compared with $\underline{H}$ by the criterion

$$|(m\,|\,\underline{V}\,|\,n)| \ll |(m\,|\,\underline{H}\,|\,m) - (n\,|\,\underline{H}\,|\,n)| \qquad (m \neq n). \qquad [29.2]$$

Using [28.25] we can write

$$E_n^{(0)}(n\,|\,\underline{S}\,|\,m) + \sum_k (n\,|\,\underline{V}\,|\,k)(k\,|\,\underline{S}\,|\,m) = (n\,|\,\underline{S}\,|\,m)E_m, \qquad [29.3]$$

where the $E_n^{(0)}$ are the eigenvalues of $\underline{H}^{(0)}$. This relation is still exact. Now, however, because of [29.2], we make an expansion: [1]

$$E_n = E_n^{(0)} + E_n^{(1)} + E_n^{(2)} + \dots, \qquad [29.4]$$

$$\underline{S} = \underline{S}^{(0)} + \underline{S}^{(1)} + \underline{S}^{(2)} + \dots, \qquad \underline{S}^{(0)} = 1. \qquad [29.5]$$

---

[1] Because of [29.2], we can first set $(n|\underline{V}|k)$ in [29.3] approximately equal to zero for $k \neq n$:

$$\{E_n^{(0)} + (n\,|\,\underline{V}\,|\,n) - E_m\}(n\,|\,\underline{S}\,|\,m) \simeq 0.$$

We can number the eigenfunctions of $\underline{H}$ such that

$$E_n \simeq E_n^{(0)} + (n\,|\,\underline{V}\,|\,n)$$

and

$$(n\,|\,\underline{S}\,|\,m) \simeq 0 \qquad (m \neq n).$$

Thus, because of $\underline{S}\,\underline{S}^\dagger = \underline{S}^\dagger\underline{S} = 1$, we have

$$(n\,|\,\underline{S}\,|\,n) \simeq 1.$$

This justifies the expansions [29.4] and [29.5].

(We have avoided introducing a parameter ε. Instead of writing

$$\underline{S} = 1 + \varepsilon \underline{S}^{(1)} + \varepsilon^2 \underline{S}^{(2)} + \varepsilon^3 \underline{S}^{(3)} + \dots \,,$$

we stipulate that the quantities $E_n^{(i)}$ and $\underline{S}^{(i)}$ in the expansions be, respectively, one order of magnitude smaller than $E_n^{(i-1)}$ and $\underline{S}^{(i-1)}$.)

a. First approximation

Substituted into [29.3], this yields

$$E_n^{(0)} \delta_{nm} + E_n^{(0)}(n \mid \underline{S}^{(1)} \mid m) + (n \mid \underline{V} \mid m)$$
$$= E_n^{(0)} \delta_{nm} + E_n^{(1)} \delta_{nm} + (n \mid \underline{S}^{(1)} \mid m) E_m^{(0)} \,,$$

$$E_n^{(0)}(n \mid \underline{S}^{(1)} \mid m) + (n \mid \underline{V} \mid m)$$
$$= E_n^{(1)} \delta_{nm} + (n \mid \underline{S}^{(1)} \mid m) E_m^{(0)} \quad [29.6]$$

to first order. In solving this equation, we want to distinguish between two cases:

1. $n \neq m$:

$$(E_n^{(0)} - E_m^{(0)})(n \mid \underline{S}^{(1)} \mid m) = -(n \mid \underline{V} \mid m)\,. \qquad [29.7]$$

In case $\underline{H}^{(0)}$ is degenerate, this equation can only be satisfied by

$$(n \mid \underline{V} \mid m) = 0 \quad \text{for} \quad E_n^{(0)} = E_m^{(0)}, \quad n \neq m; \qquad [29.8]$$

thus, if there is degeneracy, this condition must be satisfied in order that it be possible to carry out the perturbation calculation. The approximation will also be bad if the eigenvalues are very close together; $\underline{V}$ must satisfy condition [29.2] for this reason. From [29.7] we obtain

$$(n \mid \underline{S}^{(1)} \mid m) = -\frac{(n \mid \underline{V} \mid m)}{E_n^{(0)} - E_m^{(0)}}\,. \qquad [29.9]$$

2. $m = n$:

$$E_n^{(1)} = (n \mid \underline{V} \mid n)\,. \qquad [29.10]$$

Thus, in first order, the energy eigenvalues are displaced

by amounts equal to the diagonal matrix elements of the perturbation.

Since the transformation matrix $\underline{S}$ must be unitary,

$$\underline{S}^\dagger \underline{S} = 1,$$

we have

$$\underline{S}^{(1)} + \underline{S}^{(1)\dagger} = 0, \qquad [29.11]$$

in first order. This means that $(n|\underline{S}^{(1)}|n)$ is purely imaginary but otherwise completely arbitrary, corresponding to the fact that we can always make a phase transformation.

b. Second approximation

In second order, [29.3], [29.4], and [29.5] yield

$$E_n^{(0)}\delta_{nm} + E_n^{(0)}(n\,|\,\underline{S}^{(1)}\,|\,m) + E_n^{(0)}(n\,|\,\underline{S}^{(2)}\,|\,m) + (n\,|\,\underline{V}\,|\,m)$$
$$+ \sum_k (n\,|\,\underline{V}\,|\,k)(k\,|\,\underline{S}^{(1)}\,|\,m) = E_n^{(0)}\delta_{nm} + E_n^{(1)}\delta_{nn} + E_n^{(2)}\delta_{nm}$$
$$+ (n\,|\,\underline{S}^{(1)}\,|\,m)E_m^{(0)} + (n\,|\,\underline{S}^{(1)}\,|\,m)E_m^{(1)} + (n\,|\,\underline{S}^{(2)}\,|\,m)E_m^{(0)};$$

with [29.6], this becomes

$$(E_n^{(0)} - E_m^{(0)})(n\,|\,\underline{S}^{(2)}\,|\,m) + \sum_k (n\,|\,\underline{V}\,|\,k)(k\,|\,\underline{S}^{(1)}\,|\,m)$$
$$= \delta_{nm}E_n^{(2)} + (n\,|\,\underline{S}^{(1)}\,|\,m)E_m^{(1)}. \qquad [29.12]$$

Again, we distinguish between two cases:

1. $n \neq m$:

$$(n\,|\,\underline{S}^{(2)}\,|\,m) = \frac{\{(n\,|\,\underline{V}\,|\,n) - (m\,|\,\underline{V}\,|\,m)\}\,(n\,|\,\underline{V}\,|\,m)}{(E_n^{(0)} - E_m^{(0)})^2}$$
$$+ \sum_{\substack{k \\ k \neq n \\ k \neq m}} \frac{(n\,|\,\underline{V}\,|\,k)(k\,|\,\underline{V}\,|\,m)}{(E_k^{(0)} - E_m^{(0)})(E_n^{(0)} - E_m^{(0)})}. \qquad [29.13]$$

2. $n = m$:

$$E_n^{(2)} = -\sum_{\substack{k \\ k \neq n}} \frac{(n\,|\,\underline{V}\,|\,k)(k\,|\,\underline{V}\,|\,n)}{E_k^{(0)} - E_n^{(0)}} = -\sum_{\substack{k \\ k \neq n}} \frac{|(n\,|\,\underline{V}\,|\,k)|^2}{E_k^{(0)} - E_n^{(0)}} \qquad [29.14]$$

(always negative for the lowest eigenvalue).

c. Degeneracy of $\underline{H}^{(0)}$

In case $\underline{H}^{(0)}$ is degenerate, we have already seen that $\underline{V}$ must satisfy the condition [29.8] before we can perform a perturbation calculation at all. In order to achieve this with g-fold degeneracy $(E_1^{(0)} = E_2^{(0)} = \ldots = E_g^{(0)} = E_0)$, we must pick out the corresponding g-dimensional subspace and solve equation [28.25] exactly in this subspace,

$$\sum_{k=1}^{g} (n \,|\, \underline{H} \,|\, k)(k \,|\, \underline{S} \,|\, A) = E_A(n \,|\, \underline{S} \,|\, A), \qquad [29.15]$$

in which case [29.8] is satisfied. Equation [29.15] represents a set of g systems of equations (labeled with the index A), each system containing g linear, homogeneous equations (labeled with the index n) in the g unknowns $(n|\underline{S}|A)$. The matrix of the coefficients of each of these systems of equations is

$$\underline{H} - E_A \cdot 1$$
$$= \begin{Vmatrix} E_0 + V_{11} - E_A & V_{12} \cdots \cdots V_{1g} \\ V_{21} & E_0 + V_{22} - E_A & \cdot \\ \vdots & \ddots & \vdots \\ V_{g1} \cdots \cdots \cdots \cdots E_0 + V_{gg} - E_A \end{Vmatrix}, \quad [29.16]$$

and, as is well known, there are nontrivial solutions of the system only under the condition

$$\det \Vert \underline{H} - E_A \cdot 1 \Vert = 0. \qquad [29.17]$$

This is an algebraic equation of the gth degree for E_A, whose solution determines the energy eigenvalues. Once this equation has been solved, there is no further obstacle to solving [29.15].

In conclusion, we remark that this eigenvalue problem is nothing more than the famous principal axis transformation of a $g \times g$ matrix $\underline{H}$, the Hermiticity of which requires that all of the eigenvalues be real. Equation [29.17] is known as the *secular* equation because it first arose in connection with the calculation of secular perturbations of planet trajectories.

30. TIME-DEPENDENT PERTURBATION

If we have a time-dependent perturbation, then of course we are only interested in a solution of the time-dependent wave equation

$$ih \frac{\partial \psi}{\partial t} = \underline{H}_0 \psi + \underline{V}(t)\psi . \qquad [30.1]$$

We expand the solution of this equation in the usual way (see, for example, [7.7]), in this case with respect to the unperturbed eigenfunctions u_n (as proposed by Dirac):

$$\psi = \sum_n a_n(t) u_n \exp\left[-\frac{i}{h} E_n^{(0)} t\right], \qquad [30.2]$$

where

$$\underline{H}_0 u_n = E_n^{(0)} u_n . \qquad [30.3]$$

Substituting [30.2] into [30.1], and remembering [30.3], we obtain

$$ih \sum_m \frac{da_m}{dt} u_m \exp\left[-\frac{i}{h} E_m^{(0)} t\right]$$
$$= \sum_m a_m(t) \underline{V} u_m \exp\left[-\frac{i}{h} E_m^{(0)} t\right]. \qquad [30.4]$$

We multiply this equation by u_n^* and integrate over dq:

$$ih \frac{da_n}{dt} \exp\left[-\frac{i}{h} E_n^{(0)} t\right]$$
$$= \sum_m a_m(t)(n \mid \underline{V} \mid m) \exp\left[-\frac{i}{h} E_m^{(0)} t\right] \qquad [30.5]$$

or

$$ih \frac{\mathrm{d}a_n}{\mathrm{d}t} = \sum_m a_m(t)(n \,|\, \underline{V} \,|\, m) \exp\left[+ \frac{i}{h}(E_n^{(0)} - E_m^{(0)})t \right]. \qquad [30.6]$$

We seek a solution for given initial values,

$$a_n(0) = a_n^{(0)}.$$

We also expand this solution:

$$a_n(t) = a_n^{(0)} + a_n^{(1)}(t) + a_n^{(2)}(t) + \dots. \qquad [30.7]$$

Then the initial conditions are

$$a_n^{(1)}(0) = a_n^{(2)}(0) = \dots = 0. \qquad [30.8]$$

For the sake of brevity we introduce

$$(n \,|\, \Omega \,|\, m) \equiv (n \,|\, V \,|\, m) \exp\left[\frac{i}{h}(E_n^{(0)} - E_m^{(0)})t \right], \qquad [30.9]$$

and, by integrating [30.6] over t, we obtain

$$a_n^{(1)}(t) = -\frac{i}{h} \sum_m a_m^{(0)} \int_0^t (n \,|\, \Omega \,|\, m)\, \mathrm{d}t, \qquad [30.10]$$

$$a_n^{(2)}(t) = -\frac{i}{h} \sum_l \int_0^t (n \,|\, \Omega \,|\, l) a_l^{(1)}(t)\, \mathrm{d}t$$

$$= -\frac{1}{h^2} \sum_m a_m^{(0)} \sum_l \int_0^t (n \,|\, \Omega \,|\, l)(\tau)\, \mathrm{d}\tau \int_0^\tau (l \,|\, \Omega \,|\, m)(\tau')\mathrm{d}\tau'. \qquad [30.11]$$

These formulas are valid for time-dependent $\underline{V}$,[2] as well as for time-independent $\underline{V}$.[3] However, if $\underline{V}$ is constant in

---

[2] For example, they can be used to calculate the induced emission and absorption of light, in which case the perturbing operator $\underline{V}$ is given by the incident radiation field.

[3] Since we only integrate from 0, we can also assume that a constant perturbation $\underline{V}$ is "turned on" at time $t = 0$, for example, by the transition of a system to an excited state.

time, we can evaluate the integrals:

$$a_n^{(1)}(t) = -\sum_m a_m^{(0)}(n\,|\,V\,|\,m)\,\frac{\exp[(i/h)(E_n^{(0)}-E_m^{(0)})t]-1}{E_n^{(0)}-E_m^{(0)}}\,, \qquad [30.12]$$

$$a_n^{(2)}(t) = \sum_m a_m^{(0)}\sum_l (n\,|\,V\,|\,l)(l\,|\,V\,|\,m)$$

$$\times\left[\frac{\exp[(i/h)(E_n^{(0)}-E_m^{(0)})t]-1}{(E_n^{(0)}-E_m^{(0)})(E_l^{(0)}-E_m^{(0)})}\right.$$

$$\left.-\frac{\exp[(i/h)(E_n^{(0)}-E_l^{(0)})t]-1}{(E_n^{(0)}-E_l^{(0)})(E_l^{(0)}-E_m^{(0)})}\right]. \qquad [30.13]$$

The approximation is not necessarily bad if $E_n^{(0)}=E_m^{(0)}$ (resonance denominator) because, as a result of having integrated from 0 to t, the numerator vanishes then also. Thus, the term in [30.12] containing the resonance denominator is

$$-a_m^{(0)}(n\,|\,V\,|\,m)\,\frac{i}{h}\,t\,,$$

and the approximation is usable for times such that

$$|(n\,|\,V\,|\,m)t|\ll h\,.$$

On the other hand, a solution valid for a longer time can only be obtained by treating the degenerate subspace exactly.

We now treat the important special case in which one of two energy values, $E^{(0)}$, lies in the continuous spectrum. This is the case, for example, if an atom makes a transition from an excited state to the ground state and, instead of emitting a γ-quantum in the process, ejects an electron from an outer shell;[4] the free electron has a continuous spectrum (or, in the case of distant walls, very densely distributed energy eigenvalues—a quasi continuum). The

---

[4] This radiationless transition is known as the Auger effect if the excitation of the atom is a result of removing an inner electron. Details can be found, for example, in E. H. S. BURHOP, *The Auger Effect and Other Radiationless Transitions*.

initial values that we use in [30.12] are

$$a_n^{(0)} = 0 \text{ for } n \neq m, \qquad a_n^{(0)} = 1 \text{ for } n = m ; \qquad [30.14]$$

that is, we consider the ejection of an electron which is known to come from state m. We use k as the variable in the continuous spectrum, and we obtain

$$a^{(1)}(k, t) = - (k \mid V \mid m) \frac{\exp\{(i/h)[E^{(0)}(k) - E_m^{(0)}]t\} - 1}{E^{(0)}(k) - E_m^{(0)}} . \qquad [30.15]$$

We now calculate the transition probability for a transition from the state m to an arbitrary state k within the time interval from 0 to t, where state k lies in the interval

$$E_m^{(0)} - \tfrac{1}{2}\Delta E < E^0(k) < E_m^{(0)} + \tfrac{1}{2}\Delta E . \qquad [30.16]$$

The result is

$$W(k, t) = \int |a^{(1)}(k, t)|^2 \, dk$$
$$\sim |(k \mid V \mid m)|^2 \cdot \int dk \, \frac{4 \sin^2 (t/2h)[E^{(0)}(k) - E_m^{(0)}]}{[E^{(0)}(k) - E_m^{(0)}]^2} . \qquad [30.17]$$

Let $P(E)$ be the number of states on the energy shell

$$dk = P(E) \, dE . \qquad [30.18]$$

With the abbreviation

$$x = \frac{t}{2h} [E^{(0)}(k) - E_m^{(0)}], \qquad [30.19]$$

we then obtain

$$W(k, t) = P(E) |(k \mid V \mid m)|^2 \, 4 \, \frac{t}{2h} \int_{-\infty}^{+\infty} \frac{\sin^2 x}{x^2} \, dx . \qquad [30.20]$$

We justify the limits of integration $-\infty$ to $+\infty$ by the requirement

$$\frac{\Delta E \cdot t}{h} \gg 1 . \qquad [30.21]$$

The integral equals π; therefore,

$$W(k, t) = P(E) \, |\, (k \,|\, V \,|\, m) \,|^2 \, \frac{2\pi t}{h}. \qquad [30.22]$$

This formula is of especial importance, because it gives the transition probability between two completely arbitrary states. Because of the importance of the formula, Fermi called it the "Golden Rule."

Chapter 9. Angular Momentum and Spin

31. GENERAL COMMUTATION RELATIONS

The eigenvalue problem for angular momentum and the corresponding transformations have an important place in wave mechanics. We have been partially concerned with them in an exercise treating the eigenvalue problem of the symmetrical top (see Sec. 47). We restrict ourselves here to the most essential points. Consult the literature for a detailed treatment of these problems (for example, P. A. M. Dirac, *Quantum Mechanics*).

For a particle with angular momentum

$$\left.\begin{array}{l} \underline{P} \equiv (\underline{P}_1, \underline{P}_2, \underline{P}_3) = \dfrac{1}{h}(\underline{x} \times \underline{p}) \quad \begin{array}{l}\text{(measured}\\ \text{in units of } h)\end{array} \\[2mm] \underline{P}_1 = \dfrac{1}{h}(\underline{x}_2\underline{p}_3 - \underline{x}_3\underline{p}_2) = \dfrac{1}{i}\left(x_2\dfrac{\partial}{\partial\underline{x}_3} - x_3\dfrac{\partial}{\partial x_c}\right), \ \dots \\[2mm] \text{(cyclical permutations of the indices)} \end{array}\right\}, \quad [31.1]$$

we have the commutation relations

$$[\underline{P}_1, \underline{P}_2] = \underline{P}_1\underline{P}_2 - \underline{P}_2\underline{P}_1 = i\underline{P}_3, \dots \qquad [31.2]$$

(cyclical permutations of the indices).

These commutation relations also hold for an arbitrary number of particles, where

$$\begin{aligned} \underline{P}_1 &= \frac{1}{i}\sum_r\left(\underline{x}_2^{(r)}\frac{\partial}{\partial\underline{x}_3^{(r)}} - \underline{x}_3^{(r)}\frac{\partial}{\partial\underline{x}_2^{(r)}}\right) \\ &= \frac{1}{h}\sum_r(\underline{x}_2^{(r)}\underline{p}_3^{(r)} - \underline{x}_3^{(r)}\underline{p}_2^{(r)}), \ \dots \ . \end{aligned} \qquad [31.3]$$

The angular momentum operator is introduced purely formally into wave mechanics for describing the transformation properties of the wave function under rotation of the coordinate system. On the basis of the kinematical properties of the rotation group, the following more general commutation relations can be given which are independent of the particular form [31.3] of $\underline{P}$:

$$[\underline{P}_k, \underline{C}] = 0 \quad (k = 1, 2, 3), \qquad [31.4]$$

$$\left.\begin{array}{l} [\underline{P}_1, \underline{A}_1] = 0, \ldots \\ [\underline{P}_1, \underline{A}_2] = -[\underline{P}_2, \underline{A}_1] = i\underline{A}_3, \ldots \end{array}\right\}, \qquad [31.5]$$

where $\underline{C}$ is a scalar function of $\underline{p}^{(r)}$ and $\underline{x}^{(r)}$, and $\underline{A}_k$ is the kth component of a vector $\underline{A}(\underline{p}^{(r)}, \underline{x}^{(r)})$. As examples, we can take $\underline{C} = \underline{H}$ (Hamiltonian), $\underline{C} = |\underline{P}|^2$, $\underline{A} = \underline{x}^{(r)}$. The eigenvalues and matrix elements of the general operator $\underline{P}$ can be derived from these relations by purely algebraic calculation.[1]

Of course, we can also proceed analytically by expressing the differential operators [31.1] in polar coordinates (see the exercise on this topic, Sec. 45). We find that the eigenvalue equation

$$\underline{P}^2 Y = \lambda Y \qquad [31.6]$$

is identical to Eq. [18.9] for the spherical functions $Y_l(\vartheta, \varphi)$. Thus, $\underline{P}^2$ has the eigenvalues

$$\lambda = l(l+1) \quad (l = 0, 1, 2, \ldots). \qquad [31.7]$$

If we further choose the $2l+1$ linearly independent Y_l so that they are simultaneously eigenfunctions of $\underline{P}_3$, which is possible because, according to [31.4], $\underline{P}^2$ commutes with the $\underline{P}_k$,

$$\underline{P}_3 Y_l = -i \frac{\partial}{\partial \varphi} Y_l = \mu Y_l,$$

---

[1] See, for example, M. BORN and P. JORDAN, *Elementare Quantenmechanik*; P. A. M. Dirac, *Quantum Mechanics*.

then we arrive at the spherical harmonics $Y_{l,m}$ (see [18.18]), with eigenvalues

$$\mu = m; \qquad m = -l, -l+1, \ldots, +l. \qquad [31.8]$$

32. MATRIX ELEMENTS OF THE ANGULAR MOMENTUM

Here, we give a summary of the matrix elements of the angular momentum operator. As we have already emphasized, these can be derived purely algebraically from the commutation relations [31.5].

We now denote the quantum number l by j. By so doing, we mean to leave open the possibility that j assumes half-integral values, in contrast to the particular form [31.1], which, according to [31.6] and [31.7], allows only integral values of l.

Here, we write the matrix elements for fixed j in the form

$$(j, m' | \;\; | j, m) ,$$

where m labels the matrix elements subject to the condition $-j \leqslant m \leqslant +j$. The matrix then has the following appearance: [2]

$$
\begin{array}{c|cccc}
 & m = j & m = j-1 & \cdots & m = -j \\
\hline
m = j & (j, j | \; | j, j) & (j, j | \; | j, j-1) & \cdots & \cdots \\
m = j-1 & (j, j-1 | \; | j, j) & (j, j-1 | \; | j, j-1) & \cdots & \cdots \\
m = j-2 & (j, j-2 | \; | j, j) & \cdots & & \\
\vdots & \vdots & \vdots & & \\
m = -j & (j, -j | \; | j, j) & \cdots & &
\end{array}
$$

---

[2] This notation means that, for example, there are the following matrices for $j = \frac{1}{2}$:

$$\underline{P}_1 + i\underline{P}_2 = \left\| \begin{array}{cc} 0 & 1 \\ 0 & 0 \end{array} \right\|, \qquad \underline{P}_1 - i\underline{P}_2 = \left\| \begin{array}{cc} 0 & 0 \\ 1 & 0 \end{array} \right\|, \qquad \underline{P}_3 = \frac{1}{2} \left\| \begin{array}{cc} 1 & 0 \\ 0 & -1 \end{array} \right\|.$$

Corresponding to these, there are the matrices

$$\underline{P}_1 = \frac{1}{2} \left\| \begin{array}{cc} 0 & 1 \\ 1 & 0 \end{array} \right\|, \qquad \underline{P}_2 = \frac{1}{2} \left\| \begin{array}{cc} 0 & -i \\ i & 0 \end{array} \right\|.$$

In the representation in which $\underline{P}^2$ and $\underline{P}_3$ are diagonal (as is well known, commuting operators can be simultaneously diagonalized), we find

$$(j, m' | \underline{P}^2 | j, m) = j(j + 1)\delta_{mm'}, \qquad [32.1]$$

$$(j, m' | \underline{P}_3 | j, m) = m\delta_{mm'}, \qquad [32.2]$$

and, for the non-Hermitian matrices

$$\underline{P}_1 \pm i\underline{P}_2,$$

we further find

$$(j, m + 1 | \underline{P}_1 + i\underline{P}_2 | j, m) = (j, m | \underline{P}_1 - i\underline{P}_2 | j, m + 1)$$
$$= \sqrt{(j - m)(j + 1 + m)}. \qquad [32.3]$$

All other matrix elements vanish.

For every vector $\underline{A}$—in particular, for the coordinate matrices—we can derive the following general expressions from the commutation relations [31.5]:

$$
\left.
\begin{aligned}
&(j + 1, m \pm 1 | \underline{A}_1 \pm i\underline{A}_2 | j, m) \\
&\qquad = \mp (j + 1 | \underline{A} | j)\sqrt{(j \pm m + 2)(j \pm m + 1)} \\
&(j + 1, m | \underline{A}_3 | j, m) \\
&\qquad = (j + 1 | \underline{A} | j) \sqrt{(j + m + 1)(j - m + 1)} \\
&(j, m \pm 1 | \underline{A}_1 \pm i\underline{A}_2 | j, m) \\
&\qquad = (j | \underline{A} | j)\sqrt{(j \mp m)(j \pm m + 1)} \\
&(j, m | \underline{A}_3 | j, m) = (j | \underline{A} | j)m \\
&(j - 1, m \pm 1 | \underline{A}_1 \pm i\underline{A}_2 | j, m) \\
&\qquad = \pm (j - 1 | \underline{A} | j)\sqrt{(j \mp m)(j \mp m - 1)} \\
&(j - 1, m | \underline{A}_3 | j, m) = (j - 1 | \underline{A} | j)\sqrt{(j + m)(j - m)}
\end{aligned}
\right\} \qquad [32.4]
$$

For all other pairs of values j, m in the initial and final states the matrix elements vanish. The expressions $(j' | \underline{A} | j'')$ are numbers which are independent of m.

33. SPIN

In order to explain experimental observations—first in connection with anomalous Zeeman splitting—a spin was ascribed to the free electron [A-3].[3] This fact can be included in the theory by means of the general representation [32.1], [32.2], and [32.3] of the angular momentum. This representation indeed does not rest on the special definition [31.3], which cannot be applied to spin; instead, it is based solely on the relations [31.5]. On this basis, it is possible to introduce spin into the theory in a purely formal way.[4] In the relativistic treatment of the electron (Dirac equation), it turns out that the spin formalism discussed in this section is already contained in the equations and becomes evident in the limiting case of small velocities.[5]

We now want to consider the generalized description of a particle with spin. By the spin s of a particle we shall always mean an angular momentum whose magnitude (as opposed to its components) is always a fixed number. We denote the spin operators by s_k. Then, for example, the commutation relations analogous to [31.5] hold:

$$\underline{s}_1\underline{s}_2 - \underline{s}_2\underline{s}_1 = i\underline{s}_3, \ldots \tag{33.1}$$

(cyclical permutations of the indices).

Since we consider $|\underline{s}|^2$ as a fixed number, which, according to [32.1], must be of the form

$$|\underline{s}|^2 = \underline{s}_1^2 + \underline{s}_2^2 + \underline{s}_3^2 = s(s+1),$$

with s half-integral or integral, we can introduce one of the components s_k, for example, s_3, into the wave function as a new independent variable: $\psi = \psi(q, s_3, t)$. However,

[3] G. E. UHLENBECK and S. GOUDSMIT, *Naturwiss.* **13**, 953 (1925); *Nature* **117**, 264 (1926).

[4] W. PAULI, *Z. Physik* **43**, 601 (1927).

[5] P. A. M. DIRAC, *Proc. Roy. Soc. (London)* A **117**, 610 (1928); A **118**, 351 (1928).

since s_3 can only assume the values $-s, ..., +s$ (see [32.2]), we can also write

$$\psi(q, s_3; t) = \sum_\mu C_\mu(s_3)\psi_\mu(q, t) \qquad (-s \leqslant \mu \leqslant s), \quad [33.2]$$

where, for example, the $C_\mu(s_3)$ are defined by

$$C_\mu(s_3) = \begin{cases} 1 & \text{for } s_3 = \mu, \\ 0 & \text{otherwise} \end{cases} \qquad [33.3]$$

and satisfy the orthogonality relation

$$\sum_{s_3=-s}^{+s} C_\mu^*(s_3) \cdot C_{\mu'}(s_3) = \begin{cases} 1 & \text{for } \mu = \mu', \\ 0 & \text{for } \mu \neq \mu'. \end{cases} \qquad [33.4]$$

The effect of the operators $\underline{s}_k$ on a wave function is most easily seen in the matrix notation:

$$\underline{s}_k \cdot \psi_\mu(q, t) = \sum_{\mu'} \psi_{\mu'}(\mu' \,|\, \underline{s}_k \,|\, \mu), \qquad [33.5]$$

with the matrix elements given by (see [32.3])

$$\left.\begin{array}{l} (\mu \pm 1 \,|\, \underline{s}_1 \pm i\underline{s}_2 \,|\, \mu) = \sqrt{(s \mp \mu)(s + 1 \pm \mu)} \\ (\mu \,|\, \underline{s}_3 \,|\, \mu) = \mu \end{array}\right\}. \qquad [33.6]$$

Thus, for example, we have

$$\underline{s}_3 \cdot \psi_\mu = \psi_\mu \cdot \mu,$$
$$\underline{s}_1 \cdot \psi_\mu = \tfrac{1}{2}(\underline{s}_1 + i\underline{s}_2)\psi_\mu + \tfrac{1}{2}(\underline{s}_1 - i\underline{s}_2)\psi_\mu$$
$$= \tfrac{1}{2}\psi_{\mu-1}\sqrt{(s + \mu)(s + 1 - \mu)}$$
$$+ \tfrac{1}{2}\psi_{\mu+1}\sqrt{(s - \mu)(s + 1 + \mu)},$$

where the first term vanishes for $\mu = -s$ and the second term vanishes for $\mu = s$.

The most important and most fundamental application of this formalism is to the spin of the electron. Since

$$s = \tfrac{1}{2}, \quad \text{which means} \quad \mu = -\tfrac{1}{2}, +\tfrac{1}{2}, \quad [33.7]$$

everything is particularly simple:

$$|\underline{s}|^2 = \underline{s}_1^2 + \underline{s}_2^2 + \underline{s}_3^2 = s(s+1) = \tfrac{3}{4},$$

$$\underline{s}_1 + i\underline{s}_2 = \begin{Vmatrix} 0 & 1 \\ 0 & 0 \end{Vmatrix}, \qquad \underline{s}_1 - i\underline{s}_2 = \begin{Vmatrix} 0 & 0 \\ 1 & 0 \end{Vmatrix}, \qquad \underline{s}_3 = \begin{Vmatrix} \tfrac{1}{2} & 0 \\ 0 & -\tfrac{1}{2} \end{Vmatrix}.$$

With a new operator defined by

$$\boldsymbol{\sigma} = 2\underline{s}, \qquad\qquad [33.8]$$

we obtain [6]

$$\sigma_1^2 = \sigma_2^2 = \sigma_3^2 = 1 = \begin{Vmatrix} 1 & 0 \\ 0 & 1 \end{Vmatrix}, \qquad\qquad [33.9]$$

$$\sigma_1 = \begin{Vmatrix} 0 & 1 \\ 1 & 0 \end{Vmatrix}, \qquad \sigma_2 = \begin{Vmatrix} 0 & -i \\ i & 0 \end{Vmatrix}, \qquad \sigma_3 = \begin{Vmatrix} 1 & 0 \\ 0 & -1 \end{Vmatrix}. \qquad [33.10]$$

These operators obey the commutation relations

$$[\sigma_1, \sigma_2] = \sigma_1\sigma_2 - \sigma_2\sigma_1 = 2i\sigma_3, \dots \qquad [33.11]$$

(cyclical permutations of the indices).

In addition, they satisfy the relations

$$\sigma_1\sigma_2 = -\sigma_2\sigma_1 = i\sigma_3, \dots$$

(cyclical permutations of the indices)

or

$$\sigma_1\sigma_2 + \sigma_2\sigma_1 \equiv [\sigma_1, \sigma_2]_+ \equiv 0, \dots \qquad [33.12]$$

(cyclical permutations of the indices).

Because of the latter relations, the σ_k are said to anti-commute.

Now, in addition to the position coordinates and the time, the wave function of an electron will also contain the discrete spin variable s_3; the spin variable represents the additional degree of freedom of a spinning electron. In correspondence with the two eigenvalues $s_3 = +\tfrac{1}{2}$ and $-\tfrac{1}{2}$,

---

[6] The matrices σ_1, σ_2, and σ_3 are called the Pauli spin matrices [note by the students].

we divide ψ into two terms, ψ_1 and ψ_2, as in [33.2], and write

$$\psi(q, s_3; t) = \left\| \begin{matrix} \psi_1(q; t) \\ \psi_2(q; t) \end{matrix} \right\| . \qquad [33.13]$$

The normalization condition is

$$\int \psi \psi^* \, \mathrm{d}V = \int |\psi_1|^2 \mathrm{d}V + \int |\psi_2|^2 \mathrm{d}V = 1 ,$$

where $|\psi_1|^2 \mathrm{d}V$ and $|\psi_2|^2 \mathrm{d}V$ can be thought of as the probabilities that the electron has its spin respectively parallel or antiparallel to the positive z direction.

Obviously, this formalism can be immediately applied to the usual calculational methods of wave mechanics (transformation theory, perturbation theory, etc.). In general, in addition to $\underline{p}$ and $\underline{q}$, the Hamiltonian function will also contain the $\underline{s}_k$; for example, if there is a magnetic field H, the Hamiltonian will contain an additional term equal to a constant times

$$\underline{s}_1 H_1 + \underline{s}_2 H_2 + \underline{s}_3 H_3 .$$

The generalization of this formalism to N particles with spin is immediate. The wave function will now also contain the spin variable $s_3^{(a)}$ of each of the N particles ($a = 1, ..., N$), where $s_3^{(a)}$ has one of the values $-s^{(a)}, ..., +s^{(a)}$. For an electron we have simply $s_3^{(a)} = \pm \frac{1}{2}$. A total angular momentum,

$$\underline{J} = \sum_{a=1}^{N} \left\{ (\underline{x}^{(a)} \times \underline{p}^{(a)}) + h \underline{s}^{(a)} \right\} ,$$

can be defined which obeys the usual commutation relations [31.5].

34. SPINORS AND SPACE ROTATIONS

Here we want to investigate how the spin formalism that has been presented behaves with respect to rotations in space.

For this purpose we consider a 2×2 unitary matrix $\underline{S}$, for which we specifically require

$$\det \underline{S} = 1 ,$$

(From unitarity, $\underline{S}\underline{S}^\dagger = 1$, it only follows that $|\det \underline{S}|^2 = 1$; that is, there is a trivial phase factor undetermined in $\det \underline{S} = e^{i\alpha}$, where α is real.)

$$\underline{S} = \begin{Vmatrix} S_{11} & S_{12} \\ S_{21} & S_{22} \end{Vmatrix}, \qquad \det \underline{S} = S_{11} S_{22} - S_{12} S_{21} = 1 . \qquad [34.1]$$

From the condition

$$\underline{S}^{-1} = \underline{S}^\dagger , \quad \text{with} \quad \underline{S}^{-1} = \begin{Vmatrix} S_{22} & -S_{12} \\ -S_{21} & S_{11} \end{Vmatrix}, \qquad [34.2]$$

we then obtain

$$S_{22} = S_{11}^* , \qquad S_{21} = -S_{12}^* ; \qquad |S_{11}|^2 + |S_{12}|^2 = 1 .$$

With these results we can write

$$\underline{S} = \begin{Vmatrix} S_{11} & S_{12} \\ -S_{12}^* & S_{11}^* \end{Vmatrix}. \qquad [34.3]$$

We shall show that the transformation represented by the matrix $\underline{S}$ is a rotation. This representation of a rotation in space was already known before quantum mechanics.[7] The matrix $\underline{S}$ can be written, for example, as a function

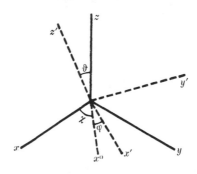

Figure 34.1

[7] See F. KLEIN and A. SOMMERFELD, Über die Theorie des Kreisels.

of the three Euler angles (Fig. 34.1):

$$S_{11} = \cos\frac{\vartheta}{2}\exp\left[\frac{i}{2}(\varphi+\chi)\right]$$
$$S_{12} = i\sin\frac{\vartheta}{2}\exp\left[\frac{i}{2}(\varphi-\chi)\right]$$

[34.4]

We now introduce a two-component mathematical form,

$$\xi = \|\xi_1, \ \xi_2\|,$$

[34.5]

called a *spinor* by Ehrenfest, which is transformed by $\underline{S}$ as follows:

$$\xi' = \xi \cdot \underline{S}, \quad \begin{aligned} \xi_1' &= \xi_1 S_{11} + \xi_2 S_{21} = \xi_1 S_{11} - \xi_2 S_{12}^* \\ \xi_2' &= \xi_1 S_{12} + \xi_2 S_{22} = \xi_1 S_{12} + \xi_2 S_{11}^* \end{aligned}$$

[34.6]

We see that our two-component wave function [33.13]

$$\psi = \left\| \begin{matrix} \psi_1 \\ \psi_2 \end{matrix} \right\|$$

[34.7]

transforms contragrediently to the ξ with respect to space rotations: [8]

$$\psi' = \underline{S}^{-1} \cdot \psi, \quad \begin{aligned} \psi_1' &= S_{22}\psi_1 - S_{12}\psi_2 = S_{11}^*\psi_1 - S_{12}\psi_2 \\ \psi_2' &= -S_{21}\psi_1 + S_{11}\psi_2 = S_{12}^*\psi_1 + S_{11}\psi_2 \end{aligned}$$

[34.8]

We note that $\|\xi_1^*, \xi_2^*\|$ transforms like $\|\xi_2, -\xi_1\|$, and that

$$\left\| \begin{matrix} \psi_1^* \\ \psi_2^* \end{matrix} \right\| \quad \text{transforms like} \quad \left\| \begin{matrix} \psi_2 \\ -\psi_1 \end{matrix} \right\|;$$

that is, the formulas remain correct if we make the following substitutions:

$$\begin{aligned} \xi_1^* &\to \xi_2 & \psi_1^* &\to \psi_2 \\ \xi_2^* &\to -\xi_1 & \psi_2^* &\to -\psi_1 \end{aligned}$$

[34.9]

---

[8] The contragredient nature of ψ with respect to ξ is characterized by the invariance of the form $(\xi\psi)$.

At this point we make the important observation that every unitary transformation, including our $\underline{S}$, leaves the form

$$N = (\xi \xi^*) = \xi_1 \xi_1^* + \xi_2 \xi_2^* \qquad [34.10]$$

or

$$\varrho = (\psi^* \psi) = \psi_1^* \psi_1 + \psi_2^* \psi_2 \qquad [34.11]$$

invariant. This can be demonstrated immediately by substituting $\xi' = \xi \underline{S}$ and $\xi'^* = \underline{S}^\dagger \xi^*$. Conversely, it can be said that this requirement, which is basic to the physical meaning of the theory, together with the requirements of unimodularity (det $\underline{S} = 1$) and linearity, determines the form of our particular transformation matrix $\underline{S}$.

Now, we want to show that the transformation $\underline{S}$ actually represents a rotation. In connection with $\boldsymbol{\sigma} = (\sigma_1, \sigma_2, \sigma_3)$ (see [33.10]), we introduce two vectors $\boldsymbol{x}$ and $\boldsymbol{d}$ ($\boldsymbol{d}$ is called the *spin density*):

$$\boldsymbol{x} = (\xi \boldsymbol{\sigma} \xi^*) , \qquad \boldsymbol{d} = (\psi^* \boldsymbol{\sigma} \psi) . \qquad [34.12]$$

This notation means, for example,

$$\boldsymbol{x} = \sum_{\alpha=1,2} \sum_{\beta=1,2} \xi_\alpha \, \sigma_{\alpha\beta} \xi_\beta^* .$$

Using

$$\sigma_1 + i\sigma_2 = 2 \begin{Vmatrix} 0 & 1 \\ 0 & 0 \end{Vmatrix}, \qquad \sigma_1 - i\sigma_2 = 2 \begin{Vmatrix} 0 & 0 \\ 1 & 0 \end{Vmatrix},$$

we obtain

$$\left. \begin{array}{ll} x_1 + ix_2 = 2\xi_1 \xi_2^* & d_1 + id_2 = 2\psi_1^* \psi_2 \\ x_1 - ix_2 = 2\xi_2 \xi_1^* & d_1 - id_2 = 2\psi_2^* \psi_1 \\ x_3 = \xi_1 \xi_1^* - \xi_2 \xi_2^* & d_3 = \psi_1^* \psi_1 - \psi_2^* \psi_2 \end{array} \right\}. \qquad [34.13]$$

We now ask how $\boldsymbol{x}$ and $\boldsymbol{d}$ transform when we transform ξ and ψ with $\underline{S}$. To answer this, we calculate the expression

$$x_1^2 + x_2^2 + x_3^2 = (x_1 + ix_2)(x_1 - ix_2) + x_3^2 .$$

Using [34.13] and [34.10], we immediately obtain

$$x_1^2 + x_2^2 + x_3^2 = N^2 = \text{constant}. \qquad [34.14]$$

Similarly, we find

$$d_1^2 + d_2^2 + d_3^2 = \varrho^2 = \text{constant}. \qquad [34.15]$$

Thus, since $\underline{S}$ leaves the forms $|\boldsymbol{x}|^2$ and $|\boldsymbol{d}|^2$ invariant, we have shown that it induces a rotation when acting on ξ and ψ. In both cases we have exactly the same rotation. As a result of the linear transformation property of such a rotation, we can also write

$$x_i' = \sum_k x_k A_{ki}, \qquad d_i' = \sum_k d_k A_{ki}. \qquad [34.16]$$

Using [34.6], [34.8], and [34.12], we immediately obtain

$$\underline{S}\, \sigma_i \underline{S}^{-1} = \sum_k \sigma_k A_{ki}. \qquad [34.17]$$

By virtue of [34.17], an $\underline{S}$ transformation, which, according to [34.4], is characterized by three real parameters, is associated with a rotation $\underline{A}$,

$$\underline{S} \rightarrow \underline{A}. \qquad [34.18]$$

It is apparent from the form of [34.17] that the inverse association is double-valued:

$$\underline{A} \rightarrow \underline{S},$$
$$\underline{A} \rightarrow -\underline{S}.$$

If, for example, ξ_1 and ξ_2 both change sign (that is, $\underline{S} = -1$), then $\boldsymbol{x}$ and $\boldsymbol{d}$ are unchanged:

$$\xi' = -\xi, \quad x' = x; \qquad \psi' = -\psi, \quad \boldsymbol{d}' = \boldsymbol{d}.$$

If

$$\underline{S}_\mathrm{I} \rightarrow \underline{A}, \qquad \underline{S}_\mathrm{II} \rightarrow \underline{B},$$

then, in the sense of matrix multiplication,

$$\underline{S}_\mathrm{I}\, \underline{S}_\mathrm{II} \rightarrow \underline{A}\, \underline{B}.$$

We show this with the help of [34.17]:

$$\underline{S}_{\mathrm{I}}\, \sigma_i\, \underline{S}_{\mathrm{I}}^{-1} = \sum_k \sigma_k A_{ki} \, ,$$

$$\underline{S}_{\mathrm{II}}\, \sigma_i\, \underline{S}_{\mathrm{II}}^{-1} = \sum_k \sigma_k B_{ki} \, ,$$

$$\underline{S}_{\mathrm{I}}(\underline{S}_{\mathrm{II}}\, \sigma_i\, \underline{S}_{\mathrm{II}}^{-1})\, \underline{S}_{\mathrm{I}}^{-1} = \sum_k \underline{S}_{\mathrm{I}}\, \sigma_k \underline{S}_{\mathrm{I}}^{-1} B_{ki} = \sum_l \sigma_l (\underline{A}\, \underline{B})_{li} \, .$$

In addition, we recall the rule

$$(\underline{S}_{\mathrm{I}}\underline{S}_{\mathrm{II}})^\dagger = \underline{S}_{\mathrm{II}}^\dagger\, \underline{S}_{\mathrm{I}}^\dagger \, , \qquad (\underline{S}_{\mathrm{I}}\underline{S}_{\mathrm{II}})^\dagger\, \underline{S}_{\mathrm{I}}\,\underline{S}_{\mathrm{II}} = 1 \, .$$

The content of this section is very closely connected with group theory, which we have not been able to introduce here. There are detailed textbooks on the relationship of wave mechanics to group theory;[9] for a deeper study of the subject, reference should be made to these books.

Remark: By dropping the restriction that $\underline{S}$ be unitary (but keeping the requirement $\det \underline{S} = 1$), the above considerations can be extended to the Lorentz group;[10] we arrive at relativistic spin theory in this way.

---

[9] B. L. van der Waerden, *Die Gruppentheoretische Methode in der Quantenmechanik*; H. Weyl, *Group Theory and Quantum Mechanics*; E. P. Wigner, *Group Theory and its Application to the Quantum Mechanics of Atomic Spectra*; W. Pauli, *Continuous Groups in Quantum Mechanics*, CERN, 1956.

[10] This is the group of Lorentz transformations, which are well known to be characterized by invariance of the form

$$x^2 + y^2 + z^2 - c^2 t^2 \, .$$

Chapter 10. Identical Particles with Spin

35. SYMMETRY CLASSES

We first consider two identical particles labeled by superscripts (1) and (2). Because the particles are identical, the associated Hamiltonian operator must be symmetric (that is, invariant under interchange of the particles). For example, for two electrons in the Coulomb field of a nucleus and in an additional magnetic field, we have

$$\underline{H} = \frac{\boldsymbol{p}^{(1)^2}}{2m} + \frac{\boldsymbol{p}^{(2)^2}}{2m} - \frac{Ze^2}{r^{(1)}} - \frac{Ze^2}{r^{(2)}}$$

$$+ \frac{e^2}{r_{12}} + \mu_0 \underline{H}(\boldsymbol{\sigma}^{(1)} + \boldsymbol{\sigma}^{(2)}); \qquad \mu_0 = \frac{ch}{2mc}. \qquad [35.1]$$

Let

$$\psi_{\mathrm{I}} = \psi(\boldsymbol{x}^{(1)}, s_3^{(1)}; \boldsymbol{x}^{(2)}, s_3^{(2)}) \qquad [35.2]$$

be a solution of the wave equation for this Hamiltonian. The additional solution

$$\psi_{\mathrm{II}} = \psi(\boldsymbol{x}^{(2)}, s_3^{(2)}; \boldsymbol{x}^{(1)}, s_3^{(1)}) = P_{12}\psi_{\mathrm{I}} \qquad [35.3]$$

follows immediately from the symmetry of the Hamiltonian operator. Here P_{12} is the operator that interchanges the two particles (the *exchange operator*). The linear combinations

$$\left. \begin{array}{l} \psi_s = \psi_{\mathrm{I}} + \psi_{\mathrm{II}} \\ P_{12}\psi_s = \psi_s \end{array} \right\} \quad \text{(symmetric solution)}, \qquad [35.4]$$

$$\left. \begin{array}{l} \psi_a = \psi_{\mathrm{I}} - \psi_{\mathrm{II}} \\ P_{12}\psi_a = - \psi_a \end{array} \right\} \quad \text{(antisymmetric solution)} \qquad [35.5]$$

have the important property that the matrix elements between symmetric and antisymmetric solutions are always exactly zero:

$$(a\,|\,\underline{H}\,|\,s) = \iint \sum_{s_3^{(1)},\,s_3^{(2)}} \psi_a^* \underline{H} \psi_s \, \mathrm{d}^3 x^{(1)} \mathrm{d}^3 x^{(2)} \,. \qquad [35.6]$$

If we interchange the particles in this equation, then the left-hand side does not change (symmetry of $\underline{H}$!), whereas the right-hand side changes sign; thus, we must have

$$(a\,|\,\underline{H}\,|\,s) = 0 \,. \qquad [35.7]$$

Therefore, a symmetric solution can never develop into an antisymmetric one, and vice versa, which is to say that there are two separate classes of particles between which no transitions are possible for any interaction: [1]

$$\left.\begin{array}{l} \left.\begin{array}{l}\text{symmetric}\\ \text{particles}\end{array}\right\} \quad \textit{bosons}, \text{ integral spin} \\[2em] \left.\begin{array}{l}\text{antisymmetric}\\ \text{particles}\end{array}\right\} \quad \textit{fermions}, \text{ half-integral spin} \end{array}\right\} \,. \qquad [35.8]$$

If there are more than two particles ($N > 2$), then, as can be shown with the group-theoretical method of representations, there is a class of solutions symmetric in all of the particles and a class antisymmetric in all of the particles:

$$P\psi_s(x^{(1)}, s_3^{(1)}; \ldots; x^{(N)}, s_3^{(N)}) = \psi_s(x^{(1)}, s_3^{(1)}; \ldots; x^{(N)}, s_3^{(N)}) \,, \qquad [35.9]$$

$$P\psi_a(x^{(1)}, s_3^{(1)}; \ldots; x^{(N)}, s_3^{(N)}) = \varepsilon_p \psi_a(x^{(1)}, s_3^{(1)}; \ldots; x^{(N)}, s_3^{(N)}) \,, \qquad [35.10]$$

where P is an arbitrary permutation, and

$\varepsilon_P = +1$ for P even,

$\varepsilon_P = -1$ for P odd (for example, for one interchange of two particles).

---

[1] The names fermion and boson refer to the fact that particles with half-integral spin obey Fermi statistics, and particles with integral spin obey Bose statistics. See W. PAULI, *Phys. Rev.* **58**, 716 (1940).

There are yet other symmetry classes, which are distinguished from those already mentioned in that they cannot be extended from N to $N+1$. Of course, if an additional particle collides with those already present (for example, in an atom), then there are transitions possible between these classes. Therefore, there can be present either only a mixture of all classes, or only the symmetrical and/or antisymmetrical classes. Experiment shows that nature oddly makes no use of the first possibility: until now only bosons and fermions have been found.

36. THE EXCLUSION PRINCIPLE

We now want to investigate the properties of the bosons and fermions somewhat more precisely. For this, we consider particles which are isolated to first approximation; that is, we neglect the forces of interaction between the particles.

Let us begin again with two uncoupled particles in states n_1 and n_2 and write their associated eigenfunctions as

$$u_{n_1}(x^{(1)}, s_3^{(1)}) \quad \text{and} \quad u_{n_2}(x^{(2)}, s_3^{(2)}) \ .$$

We know that the wave function for uncoupled particles is the product of the individual wave functions:

$$u_s = \{u_{n_1}(x^{(1)}, s_3^{(1)}) \cdot u_{n_2}(x^{(2)}, s_3^{(2)})$$
$$+ \ u_{n_2}(x^{(1)}, s_3^{(1)}) \cdot u_{n_1}(x^{(2)}, s_3^{(2)})\} \cdot C_{n_1 n_2} , \qquad [36.1]$$

$$u_a = \{u_{n_1}(x^{(1)}, s_3^{(1)}) \cdot u_{n_2}(x^{(2)}, s_3^{(2)})$$
$$- \ u_{n_2}(x^{(1)}, s_3^{(1)}) \cdot u_{n_1}(x^{(2)}, s_3^{(2)})\} \cdot C_{n_1 n_2} , \qquad [36.2]$$

$$C_{n_1 n_2} = \begin{cases} 1/2 & \text{for } n_1 = n_2 , \\ 1/\sqrt{2} & \text{for } n_1 \neq n_2 . \end{cases} \qquad [36.3]$$

If the two functions are identical, we obtain from this

$$u_s = u_{n_1}(x^{(1)}, s_3^{(1)}) \cdot u_{n_1}(x^{(2)}, s_3^{(2)}) , \qquad [36.4]$$

$$u_a \equiv 0 . \qquad [36.5]$$

For fermions, the state for which both particles are in the same state does not exist! That is exactly the prediction of the *exclusion principle*, which was postulated for electrons even before the invention of wave mechanics.[2] We now know that it also holds for all other fermions (for example, protons and neutrons).

From a superficial consideration of the exclusion principle, it might be thought that a sort of action-at-a-distance is being postulated, as a result of which even two widely separated particles are aware of one another ("sign a contract"). However, this is not so, because the exclusion principle is only valid as long as the wave packets of the two particles overlap (Fig. 36.1). If the wave packets do

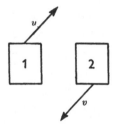

Separated wave packets. Particles can be followed individually. Exclusion principle inoperative.

Overlapping wave packets. Particles cannot be followed individually. Exclusion principle operative.

Figure 36.1

not overlap, then, everywhere in space, we have

$$u_{n_1}(\boldsymbol{x}^{(1)}, s_3^{(1)}) \cdot u_{n_2}(\boldsymbol{x}^{(1)}, s_3^{(1)}) \equiv 0 , \qquad [36.6]$$

or, for nonstationary states,

$$u(\boldsymbol{x}^{(1)}, s_3^{(1)}, t) \cdot v(\boldsymbol{x}^{(1)}, s_3^{(1)}, t) = 0 . \qquad [36.7]$$

In that case, since according to [36.6] we can have $u_{n_1} = u_{n_2}$ only if $u_{n_1} = 0$, the above symmetrization is of no consequence.

[2] W. PAULI, Z. Physik **31**, 765 (1925).

The generalization to many particles must simply be calculated combinatorially. When we symmetrize we must sum over all $N!$ permutations:

$$u_s = c \cdot \sum_P P\{u_{n_i}(x^{(j)}, s_3^{(j)})\} \qquad (i, j = 1, 2, ..., N), \qquad [36.8]$$

$$u_a = c \cdot \sum_P \varepsilon_P P\{u_{n_i}(x^{(j)}, s_3^{(j)})\}$$

$$= c \cdot \begin{vmatrix} u_{n_1}(x^{(1)}, s_3^{(1)}) & u_{n_1}(x^{(2)}, s_3^{(2)}) & \cdots & u_{n_1}(x^{(N)}, s_3^{(N)}) \\ u_{n_2}(x^{(1)}, s_3^{(1)}) & \cdots & \cdots & \vdots \\ \vdots & & & \vdots \\ u_{n_N}(x^{(1)}, s_3^{(1)}) & \cdots & \cdots & u_{n_N}(x^{(N)}, s_3^{(N)}) \end{vmatrix}. \qquad [36.9]$$

If any two of the functions u_{n_i} are equal, this determinant is identically zero. Again, this is exactly the content of the exclusion principle for an arbitrary number of fermions whose wave functions overlap.

37. THE HELIUM ATOM

We obtain the helium spectrum to good approximation with the assumption that the Hamiltonian operator is the sum of two terms, one symmetric in the space coordinates and one symmetric in the spin coordinates. This case, in which the spin-orbit interaction is small compared to the Coulomb interaction, is also known as Russel-Saunders coupling.

Let us first consider the spins of the two electrons alone. For the spin eigenfunctions we take (see [33.3])

$$C_+(s_3) = \begin{cases} 1 \text{ for } s_3 = +\tfrac{1}{2} \\ 0 \text{ for } s_3 = -\tfrac{1}{2} \end{cases}$$

$$C_-(s_3) = \begin{cases} 0 \text{ for } s_3 = +\tfrac{1}{2} \\ 1 \text{ for } s_3 = -\tfrac{1}{2} \end{cases}. \qquad [37.1]$$

From this we can form only one antisymmetric combination:

$$C^a(s_3^{(1)}, s_3^{(2)}) = \frac{1}{\sqrt{2}} \left(C_+(s_3^{(1)}) \cdot C_-(s_3^{(2)}) - C_-(s_3^{(1)}) \cdot C_+(s_3^{(2)}) \right). \qquad [37.2]$$

Using [33.5] and [33.8] it is easy to verify that

$$(\sigma_k^{(1)} + \sigma_k^{(2)}) \cdot C^a(s_3^{(1)}, s_3^{(2)}) \equiv 0 . \qquad [37.3]$$

Thus, we obtain only one eigenvalue (that is, zero); the antisymmetric state is a singlet.

On the other hand, we have three possibilities for forming a symmetric combination:

$$\left.\begin{aligned}
C_1^s(s_3^{(1)}, s_3^{(2)}) &= C_+(s_3^{(1)}) \cdot C_+(s_3^{(2)}) \\
C_0^s(s_3^{(1)}, s_3^{(2)}) &= \frac{1}{\sqrt{2}} [C_+(s_3^{(1)}) \cdot C_-(s_3^{(2)}) \\
&\qquad + C_-(s_3^{(1)}) \cdot C_+(s_3^{(2)})] \\
C_{-1}^s(s_3^{(1)}, s_3^{(2)}) &= C_-(s_3^{(1)}) \cdot C_-(s_3^{(2)})
\end{aligned}\right\} . \qquad [37.4]$$

The relations

$$\left.\begin{aligned}
\tfrac{1}{2}(\sigma_3^{(1)} + \sigma_3^{(2)}) C_1^s &= 1 \cdot C_1^s, & m_s &= 1 \\
\tfrac{1}{2}(\sigma_3^{(1)} + \sigma_3^{(2)}) C_0^s &= 0, & m_s &= 0 \\
\tfrac{1}{2}(\sigma_3^{(1)} + \sigma_3^{(2)}) C_{-1}^s &= (-1) \cdot C_{-1}^s, & m_s &= -1
\end{aligned}\right\} \qquad [37.5]$$

hold for these combinations; that is, the symmetric state is a triplet. We have characterized the three triplet states by a quantum number m_s for the s_3 component of the total spin. We also have

$$\{\tfrac{1}{2}(\boldsymbol{\sigma}^{(1)} + \boldsymbol{\sigma}^{(2)})\}^2 C_{m_s}^s = \tfrac{1}{2}\{3 + \boldsymbol{\sigma}^{(1)} \cdot \boldsymbol{\sigma}^{(2)}\} C_{m_s}^s = 2 C_{m_s}^s . \qquad [37.6]$$

Now, we must correctly combine the symmetry classes of the spin eigenfunctions with the symmetry classes of the space functions

$$u^s = \frac{1}{\sqrt{2}}\{u(\boldsymbol{x}^{(1)}) \cdot v(\boldsymbol{x}^{(2)}) + u(\boldsymbol{x}^{(2)}) \cdot v(\boldsymbol{x}^{(1)})\}, \qquad [37.7]$$

$$u^a = \frac{1}{\sqrt{2}}\{u(\boldsymbol{x}^{(1)}) \cdot v(\boldsymbol{x}^{(2)}) - u(\boldsymbol{x}^{(2)}) \cdot v(\boldsymbol{x}^{(1)})\}. \qquad [37.8]$$

The total wave function for the electrons (fermions!) must

be antisymmetric (with respect to simultaneous interchange of space and spin). There are two possibilities for
doing this:

$$U^a(x^{(1)}, s_3^{(1)}; x^{(2)}, s_3^{(2)}) = u^s(x^{(1)}, x^{(2)}) \cdot C^a(s_3^{(1)}, s_3^{(2)}) \Big\}, \quad [37.9]$$
$$\text{singlet}$$

$$V^a(x^{(1)}, s_3^{(1)}; x^{(2)}, s_3^{(2)}) = u^a(x^{(1)}, x^{(2)}) \cdot C^s_{m_s}(s_3^{(1)}, s_3^{(2)}) \Big|$$
$$\text{triplet} \Big\} \cdot \quad [37.10]$$
$$m_s = +1, 0, -1.$$

Therefore, the states of the helium atom fall into two
classes:

$$\text{singlet:} \quad \text{para helium},$$

$$\text{triplet:} \quad \text{ortho helium}.$$

In the ground state,

$$u = v, \text{ so that } u^a = 0, \text{ or } V^a = 0, \quad [37.11]$$

which means that there is only a singlet term.

In the approximation considered here, in which the Hamiltonian operator is symmetric in the space coordinates
by themselves, it follows from [35.7] that ortho and para
states do not combine. This is a result typical of wave
mechanics that was not understandable earlier.

In order to determine the energy eigenvalues, we consider the Coulomb interaction between the electrons,

$$V(x^{(1)}, x^{(2)}) = V(x^{(2)}, x^{(1)}) = \frac{e^2}{r_{12}}, \quad [37.12]$$

as a perturbation (an exact solution has not yet been
found). With the Coulomb integral

$$J_0 = \int |u(x^{(1)})|^2 |v(x^{(2)})|^2 V(x^{(1)}, x^{(2)}) \, d^3 x^{(1)} d^3 x^{(2)} \quad [37.13]$$

and the exchange integral [3]

$$J_1 = \int u^*(x^{(1)}) \cdot u(x^{(2)}) \cdot v(x^{(1)}) \cdot v^*(x^{(2)}) \cdot V(x^{(1)}, x^{(2)}) \, d^3x^{(1)} \, d^3x^{(2)},$$

[37.14]

we find the following shifts in the energy eigenvalues using [29.10]:

$$\Delta E_{sing} = J_0 + J_1 , \qquad\qquad [37.15]$$

$$\Delta E_{trip} = J_0 - J_1 . \qquad\qquad [37.16]$$

The difference $2J_1$ between the triplet and singlet term is large compared with the spin-orbit coupling (splitting of the triplet state); it is of the order of magnitude of the electrostatic energy.

Better approximations can already be found in the earliest articles about the helium spectrum.[4]

---

[3] The frequency of electron exchange is connected with the exchange integral as follows:

If we consider the state

$$u(x^{(1)}, x^{(2)}, t = 0) = \frac{1}{\sqrt{2}} \left(u^s(x^{(1)}, x^{(2)}) + u^a(x^{(1)}, x^{(2)}) \right),$$

at time t we have

$$u(x^{(1)}, x^{(2)}, t) = \frac{1}{\sqrt{2}} \left\{ u^s(x^{(1)}, x^{(2)}) \exp\left[-\frac{i}{h} E^s \cdot t \right] + u^a(x^{(1)}, x^{(2)}) \exp\left[-\frac{i}{h} E^a \cdot t \right] \right\}.$$

With

$$E^s = E^a + 2J_1$$

we obtain

$$u(x^{(1)}, x^{(2)}, t) = \frac{1}{\sqrt{2}} \left\{ u^s(x^{(1)}, x^{(2)}) \exp\left[-\frac{2i}{h} J_1 t \right] + u^a(x^{(1)}, x^{(2)}) \right\} \exp\left[-\frac{i}{h} E^a \cdot t \right],$$

from which follows

$$\left| u\left(x^{(1)}, x^{(2)}, t = \frac{\pi h}{2J_1} \right) \right| = \frac{1}{\sqrt{2}} |u^s - u^a| = |u(x^{(2)}, x^{(1)}, t = 0)|.$$

The electrons have exchanged their positions in time $t = \pi h / 2J_1$; the corresponding angular frequency is

$$\omega = \frac{2J_1}{h}.$$

[4] W. HEISENBERG, Z. Physik 38, 411 (1926) and 39, 499 (1927).

38. COLLISION OF TWO IDENTICAL PARTICLES: MOTT'S THEORY [5]

We treat the collision of two identical fermions (charge e, spin $\tfrac{1}{2}$). Let x be the relative separation of the two particles:

$$x = x^{(1)} - x^{(2)} . \qquad [38.1]$$

Earlier we calculated the wave function for scattering by the Coulomb potential (see, for example, [19.8] and the remark on p. 106). Here, we can therefore write

$$u(x) = P + S \cdot f(\theta) , \qquad [38.2]$$

$$P = \exp[ikz + i\gamma \log k(r - z)] , \qquad [38.3]$$

$$S = \frac{1}{r} \exp[ikr - i\gamma \log kr] , \qquad [38.4]$$

$$\gamma = \frac{1}{ka_0} = \frac{e^2}{hv} , \qquad [38.5]$$

$$f(\theta) = - \frac{e^2}{2mv^2 \sin^2 \tfrac{1}{2}\theta}$$
$$\cdot \exp[- i\gamma \log (1 - \cos \theta) - 2i\sigma(0, - 1/\gamma)] \qquad [38.6]$$

($m =$ reduced mass). Because the particles are identical, we cannot distinguish between θ and $\pi - \theta$ or between x

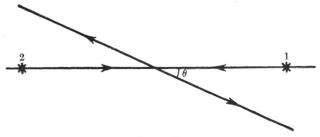

Figure 38.1

and $-x$. Classically we would add the intensities at θ and $\pi - \theta$; however, wave mechanically we must write, for un-

[5] N. F. MOTT, *Proc. Roy. Soc. (London)* A **126**, 259 (1930).

polarized particles,

$$W(x) = \tfrac{1}{4}\left\{3\,|u(x) - u(-x)|^2 + 1\,|u(x) + u(-x)|^2\right\}. \qquad [38.7]$$

$$\underset{\text{triplet}}{} \quad \underset{\text{singlet}}{}$$

The weighting factor of 3 arises because of summation over the unobserved spin orientations, $m_s = -1,\ 0,\ +1$, in the final state. Equation [38.7] contains an interference term which is typical of wave mechanics:

$$W(x) = |u(x)|^2 + |u(-x)|^2 - \tfrac{1}{2}\left\{u(x)u^*(-x) + u^*(x)u(-x)\right\}.$$

The particles with antiparallel spin orientations are still distinguishable after the collision [A-4].

From the singlet and triplet scattering amplitudes

$$|f(\theta) \pm f(\pi - \theta)|^2 = |f(\theta)|^2 + |f(\pi - \theta)|^2$$
$$\pm \left(f(\theta)f^*(\pi - \theta) + f^*(\theta)f(\pi - \theta)\right), \qquad [38.8]$$
$$\underset{\text{interference term}}{}$$

we obtain

$$dQ = \tfrac{1}{4}\left\{3\,|f(\theta) - f(\pi - \theta)|^2 + |f(\theta) + f(\pi - \theta)|^2\right\}d\Omega$$

$$= \frac{e^4}{4m^2v^4}\left\{\frac{1}{\sin^4 \tfrac{1}{2}\theta} + \frac{1}{\cos^4 \tfrac{1}{2}\theta} + \frac{1}{\sin^2 \tfrac{1}{2}\theta \cos^2 \tfrac{1}{2}\theta}\right.$$
$$\left. \times \cos\left(\gamma \log \frac{1 - \cos\theta}{1 + \cos\theta}\right)\right\}d\Omega \qquad [38.9]$$

for the differential cross section. The interference term in this formula is characteristic of wave mechanics. The interference maxima come closer together as the velocity is decreased. The special case $\theta = \pi/2$ is peculiar, in that all of the four terms in formula [38.8] are then equal.

For two identical bosons with spin zero (for example, α particles), [38.7] and [38.9] are to be replaced by

$$W(x) = |u(x) + u(-x)|^2$$

and

$$dQ = |f(\theta) + f(\pi - \theta)|^2 d\Omega.$$

According to [38.8], there is an interference term here also. The interference term was experimentally demonstrated with, for example, electrons, protons, and α particles; thus, the fermion or boson character of these particles was verified.

39. THE STATISTICS OF NUCLEAR SPINS

With a simple and elementary method we shall now show that the nuclear spin [A-3] has an important influence on the statistics of rotational states (for example, in a diatomic gas).

We consider a molecule X_2 that is composed of two identical atoms X. The eigenfunctions of such a rotator ("dumbbell") are known to be spherical harmonics (see Sec. 31), and the eigenvalues are

$$E_{\text{rot}} = \frac{h^2}{2A} l(l+1) ,$$

where A is the moment of inertia of the molecule.[6] For even l, the spherical harmonics are even:

$$Y_l(\vartheta, \varphi) = (-1)^l Y_l(\vartheta', \varphi'); \qquad \vartheta' = \pi - \vartheta , \quad \varphi' = \pi + \varphi .$$

For zero nuclear spin we have Bose statistics, and only these symmetrical states occur. However, if the nuclear spin I is different from zero, the situation is different. A nuclear spin I has the $2I+1$ orientation possibilities m_I,

$$-I \leqslant m_I \leqslant +I .$$

Therefore, the molecule has $(2I+1)^2$ possible nuclear spin states $m_I^{(1)}$, $m_I^{(2)}$, which fall into three classes:

$$C_{m_I}^{(1)} \cdot C_{m_I}^{(2)} \qquad\qquad 2I+1 \text{ states} \left.\begin{array}{c} \\ \\ \end{array}\right| (I+1)(2I+1)$$

$$C_{m_I'}^{(1)} \cdot C_{m_I''}^{(2)} + C_{m_I'}^{(2)} \cdot C_{m_I''}^{(1)} \quad m_I' \neq m_I'' \quad I(2I+1) \text{ states} \left|\quad \text{in all}\right.$$

$$C_{m_I'}^{(1)} \cdot C_{m_I''}^{(2)} - C_{m_I'}^{(2)} \cdot C_{m_I''}^{(1)} \quad m_I' \neq m_I'' \quad I(2I+1) \text{ states} .$$

[6] The separations of these rotational states ($l = 0, 1, 2, \ldots$) are very small in comparison with the excitation energy of a molecular vibration. Therefore, to each vibrational level belongs an entire band of rotational levels, and this is the reason for the typical band spectra of molecules.

As before, we must correctly combine these spin functions with the space functions:

a. I integral, bosons: (symmetric in spin + space)

$$Q = \frac{\text{number of states with even } l}{\text{number of states with odd } l} = \frac{I+1}{I} .$$

b. I half-integral, fermions: (antisymmetric in spin + space)

$$Q = \frac{\text{number of states with even } l}{\text{number of states with odd } l} = \frac{I}{I+1} .$$

Here Q is exactly the intensity ratio of adjacent lines ($l = 0, 1, 2, ...$) in a band and, therefore, it can be deduced from molecular spectra. From a knowledge of Q, the value of I and the symmetry character of the nucleus can be determined.

Historically, this played a role in determining whether the atomic nucleus is composed of protons and electrons or of protons and neutrons.[7]

---

[7] See W. PAULI, "Zur älteren und neueren Geschichte des Neutrinos" in *Aufsätze und Vorträge über Physik und Erkenntnistheorie* (Vieweg, Braunschweig, 1961).

Exercises

40. FUNDAMENTAL SOLUTION FOR INTERVAL

1. By the following two methods, determine the fundamental solution for a particle that can move freely between two parallel walls which are separated by a distance L:

a. By appropriate superposition of eigensolutions

$$\psi_n(x, t) = u_n(x) \exp\left[-\frac{i}{h} E_n t\right],$$

taking the completeness relation

$$\sum_n u_n(x) u_n^*(x') = \delta(x - x')$$

into account.

b. Using the fundamental solution for a completely free particle and using the method of images to satisfy the boundary conditions.

Show also that the expressions obtained with the two methods are identical. For this, use a property of the ϑ function defined by [1]

$$\vartheta_3(z \,|\, \tau) = \sum_{n=-\infty}^{+\infty} \exp[2niz] \exp[i\pi\tau n^2]$$

$$= 1 + 2 \sum_{n=1}^{\infty} \cos 2nz \cdot \exp[i\pi\tau n^2],$$

[1] See E. T. WHITTAKER and G. N. WATSON, *A Course of Modern Analysis* (Cambridge University Press, New York, 1962), pp. 462 ff.

the needed property being

$$\vartheta_3(z \mid \tau) = (-i\tau)^{-\frac{1}{2}} \exp\left[\frac{z^2}{i\pi\tau}\right] \cdot \vartheta_3\left(\frac{z}{\tau}\middle| -\frac{1}{\tau}\right).$$

2. Using the fundamental solution $K(x, x', t)$ found in the previous part, determine the motion of a wave packet according to

$$\psi(x, t) = \int_0^L dx' f(x') K(x, x', t) , \qquad [40.1]$$

assuming that the wave packet has the form $f(x)$ for $t = 0$. Choose for $f(x)$ the series obtained by reflection of the Gaussian distribution

$$f(x - x_0) = (\sigma_0 \sqrt{2\pi})^{-\frac{1}{2}} \cdot \exp\left[\left\{-\frac{(x - x_0)^2}{4\sigma_0^2} + \frac{i}{h}(x - x_0) m v_0\right\}\right].$$

Show that [40.1] can also be represented by

$$\psi(x, t) = \sum_{n=-\infty}^{+\infty} \{\psi_0(2nL + (x - x_0), t) \\ - \psi_0(2nL - (x - x_0), t)\}, \qquad [40.2]$$

where $\psi_0(x - x_0, t)$ is the solution for a completely free particle which belongs to the initial distribution f. Using [40.2], discuss the interferences in the probability distribution, $P(x, t) = \psi\psi^*$, by separating P into a classical term,

$$P_c = \sum_n \{|\psi_n^+|^2 + |\psi_n^-|^2\} ,$$

and an interference term,

$$P_i = \sum_n \{\psi_n^+ \psi_n^{-*} + \psi_n^+ \psi_{n+1}^{-*} + \psi_n^- \psi_n^{+*} + \psi_n^- \psi_{n+1}^{+*}\},$$

where $\psi_n^\pm$ is defined by

$$\psi_n^\pm \equiv \psi_0(2nL \pm (x - x_0), t) .$$

Show that the remainder is negligible for times that are not too large ($\sigma_0^2 + h^2 t^2 / 4m^2 \sigma_0^2 \ll L^2$). In addition, represent P_c with ϑ_3 functions and show, by means of the property of the ϑ_3 function given in the first part, that P_c goes to the uniform distribution $1/L$ for $t \to \infty$ [A-5].

41. BOUND STATES AND TUNNEL EFFECT

1. From the condition that the wave function, along with its first derivative, be continuous at the discontinuities of the potential, determine the energy eigenvalues (bound states):

a. for a one-dimensional potential well (Fig. 41.1*a*);
b. for a three-dimensional, spherically symmetric well (Fig. 41.1*b*), under the assumption that the wave function too is spherically symmetric (*S* state).

Discuss the number of eigenvalues as a function of the dimensions of the well.

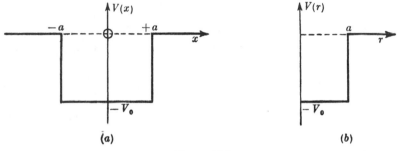

(a)   (b)

Figure 41.1

2. With the same condition as in the first part, calculate the transmission coefficient $|u_2/u_1|^2$ for a rectangular barrier as a function of the energy E (> 0). This is an example of the tunnel effect (Fig. 41.2).

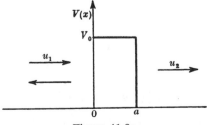

Figure 41.2

42. KRONIG-PENNEY POTENTIAL

Derive the equation which determines the energy eigenvalues E for a periodic rectangular potential (the Kronig-Penney potential). Use the condition that the wave function, along with its first derivative, is continuous at the

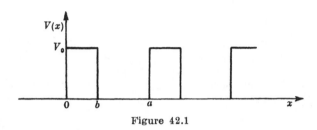

Figure 42.1

discontinuities of the potential, as well as the following general property of the eigenfunctions of a periodic potential:

$$\psi(x) = \exp[ikx] u_k(x) ; \qquad u_k(x) \text{ periodic} .$$

Discuss the behavior of $E(k)$ in the limiting case $b \to 0$, $ab(mV_0/\hbar^2) = P = $ constant.

43. SPHERICAL HARMONICS

The spherical function $Y_l(\vartheta, \varphi)$, which yields a harmonic polynomial of degree l when it is multiplied by r^l, satisfies the differential equation

$$\frac{1}{\sin\vartheta} \cdot \frac{\partial}{\partial\vartheta}\left(\sin\vartheta \frac{\partial Y_l}{\partial\vartheta}\right) + \frac{1}{\sin^2\vartheta} \cdot \frac{\partial^2 Y_l}{\partial\varphi^2} + l(l+1) Y_l = 0 .$$

It can be expanded in terms of the $2l+1$ linearly independent functions

$$Y_{l,m}(\vartheta, \varphi) = P_l^m(x) e^{im\varphi} ,$$

where $x = \cos\vartheta$ and $-l \leqslant m \leqslant +l$; the $Y_{l,m}$ satisfy the differential equation

$$(1-x^2)y'' - 2xy' + \left\{l(l+1) - \frac{m^2}{1-x^2}\right\}y = 0 .$$

There is only one solution for a given value of m^2 that is finite at $x = +1$ and $x = -1$.

1. With the aid of the substitution

$$y = (1-x^2)^{m/2}v ,$$

and starting with the cases $m = \pm l$, show that the differential equation for P_l^m is solved by

$$P_l^m(x) = (1-x^2)^{m/2} \frac{(-1)^l}{2^l l!} \frac{\mathrm{d}^{m+l}}{\mathrm{d}x^{m+l}} (1-x^2)^l \qquad [43.1]$$

(the numerical factor here is the conventional one). Also, by applying this formula for both positive and negative m, prove the relation

$$P_l^m = C_{l,m} P_l^{-m}; \quad \text{where} \quad C_{l,m} = (-1)^m \frac{(l+m)!}{(l-m)!} .$$

The value of $C_{l,m}$ can be verified, for example, by comparing with the coefficient of the highest power of x in [43.1].

2. Starting with the identity

$$\frac{\mathrm{d}^2}{\mathrm{d}x^2}(1-x^2)^{\lambda+1} + 2(\lambda+1)\{(2\lambda+1)(1-x^2)^\lambda$$
$$- 2\lambda(1-x^2)^{\lambda-1}\} = 0 , \qquad [43.2]$$

where λ is an arbitrary real number, derive the recursion formula

$$P_{l+1}^{m+1} - P_{l-1}^{m+1} = (2l+1)\sqrt{1-x^2}\,P_l^m \qquad [43.3]$$

by differentiating $l+m$ times. Using

$$\frac{\mathrm{d}^{l+m+1}}{\mathrm{d}x^{l+m+1}}(1-x^2)^{l+1} = -2(l+1)$$
$$\times \left\{x \frac{\mathrm{d}^{l+m}}{\mathrm{d}x^{l+m}}(1-x^2)^l + (l+m)\frac{\mathrm{d}^{l+m-1}}{\mathrm{d}x^{l+m-1}}(1-x^2)^l\right\},$$

verify the additional identity

$$x P_l^m + (l + m)\sqrt{1 - x^2}\, P_l^{m-1} - P_{l+1}^m = 0 \,,$$

which can be brought into the form

$$(2l + 1)x P_l^m = (l - m + 1)P_{l+1}^m + (l + m)P_{l-1}^m \qquad [43.4]$$

by applying [43.3].

3. Calculate the normalization integral

$$\int_{-1}^{+1}\{P_l^m(x)\}^2\, \mathrm{d}x = N_l^m \,.$$

For this purpose, derive a recursion formula from [43.4] which connects N_{l+1}^m with N_l^m. For the case $m = l$, there is an additional recursion formula that follows from [43.2]. The result is

$$N_l^m = \frac{(l + m)!}{(l - m)!}\frac{2}{2l + 1} \,. \qquad [43.5]$$

4. Using the normalized spherical harmonics,

$$\overline{Y}_{l,\,m}(\vartheta, \varphi) = \frac{1}{\sqrt{2\pi}}\frac{1}{\sqrt{N_l^m}}\, P_l^m(\cos\vartheta)\, e^{im\varphi} \,,$$

derive the matrix elements

$$(l', m + 1|e_1 + ie_2|l, m) \quad \text{and} \quad (l', m|e_3|l, m)$$

of the unit vector e whose components are

$$e_1 + ie_2 = \sin\vartheta\, e^{i\varphi} \quad \text{and} \quad e_3 = \cos\vartheta \,.$$

They are different from zero only when $l' = l + 1$ or $l' = l - 1$.

44. FUNDAMENTAL SOLUTION FOR HARMONIC OSCILLATOR

Determine the fundamental solution (where $\tau = \omega_0 t$)

$$K(x, x', t) = \sum_n h_n^*(x')h_n(x)\exp[-i(n + \tfrac{1}{2})\tau] \qquad [44.1]$$

of the harmonic oscillator by evaluating the sum over n in closed form. For this purpose it is convenient to use the representation

$$\exp[-x^2] = \frac{1}{\sqrt{\pi}} \int_{-\infty}^{+\infty} \exp[-v^2 + 2ivx] \, dv, \qquad [44.2]$$

from which

$$H_n(x) = (-1)^n \exp[x^2] \left(\frac{d}{dx}\right)^n \exp[-x^2]$$

$$= (-1)^n \exp[x^2] \frac{1}{\sqrt{\pi}} \int_{-\infty}^{+\infty} (2iv)^n \exp[-v^2 + 2ivx] \, dv \quad [44.3]$$

follows for the Hermite polynomials; thus,

$$h_n(x) = \frac{1}{\sqrt{2^n n! \sqrt{\pi}}} (-1)^n \exp\left[+\frac{x^2}{2}\right] \frac{1}{\sqrt{\pi}}$$

$$\times \int_{-\infty}^{+\infty} (2iv)^n \exp[-v^2 + 2ivx] \, dv. \qquad [44.4]$$

Substitution of this representation into [44.1] yields a double integral for $K(x, x', t)$, in the integrand of which the summation over n can be carried out. The double integral can then be easily evaluated by means of [44.2]. The result is

$$K(x, x', t) = \frac{1}{\sqrt{2\pi i \sin \tau}} \exp\left[i \frac{(x^2 + x'^2) \cos \tau - 2xx'}{2 \sin \tau}\right].$$

The completeness relation

$$K(x, x', 0) = \sum_n h_n^*(x') h_n(x) = \delta(x - x')$$

also follows directly from the integral obtained for K.

45. ANGULAR MOMENTUM

1. Show that the components of angular momentum can be represented in polar coordinates by

$$P_1 + iP_2 = e^{i\varphi}\left(\frac{\partial}{\partial\vartheta} + i\frac{\cos\vartheta}{\sin\vartheta}\cdot\frac{\partial}{\partial\varphi}\right),$$

$$P_1 - iP_2 = e^{-i\varphi}\left(-\frac{\partial}{\partial\vartheta} + i\frac{\cos\vartheta}{\sin\vartheta}\cdot\frac{\partial}{\partial\varphi}\right).$$

In addition, show that application of these operators to $Y_{l,m}(\vartheta, \varphi) = P_l^m e^{im\varphi}$ yields

$$(P_1 + iP_2)\,Y_{l,m}(\vartheta, \varphi) = -\,Y_{l,m+1}(\vartheta, \varphi),$$

$$(P_1 - iP_2)\,Y_{l,m}(\vartheta, \varphi) = -(l + m)(l - m + 1)\,Y_{l,m-1}(\vartheta, \varphi).$$

The equation (see Sec. 43)

$$P_l^m = (-1)^m \frac{(l+m)!}{(l-m)!}\,P_l^{-m} \qquad\qquad [45.1]$$

can be used in proving the second formula.

2. The matrix elements

$$(l,\, m + 1\,|\,P_1 + iP_2\,|\,l,\, m) \quad \text{and} \quad (l,\, m - 1\,|\,P_1 - iP_2\,|\,l,\, m)$$

are to be calculated by using the value of the normalization integral $\int \{P_l^m(x)\}^2 dx$ (Sec. 43). Verify that $P_1 - iP_2$ is Hermitian conjugate to $P_1 + iP_2$.

3. By means of the substitution $t \to -(1 - x^2)/t$, verify [45.1] with the help of the complex integral representation

$$P_l^m(x) = (1 - x^2)^{m/2}\frac{(-1)^l}{2^l l!}\cdot\frac{(l+m)!}{2\pi i}\oint\frac{\{1 - (x+t)^2\}^l}{t^{l+m+1}}\,dt\,.$$

Here, it is essential that not only $l \pm m$ be integral but that l be integral also. [Regularity of the integrand for $t = 1 - x$ and $t = -(1 + x)$.]

46. PARTIAL WAVES

1. If the potential $V_l(r)$ in the wave equation

$$\frac{1}{r}\cdot\frac{\mathrm{d}^2}{\mathrm{d}r^2}(rv_k) + k^2 v_k - \frac{2m}{\hbar^2}V_l(r)\cdot v_k = 0;$$

falls off faster than $1/r$ for large r, then the asymptotic behavior of the wave function v_k for large r is given by

$$v_k(r) \sim \frac{C}{r}\sin\left(kr - l\frac{\pi}{2} + \delta_l(k)\right); \qquad [46.1]$$

in the force-free case, $(2m/\hbar^2)V_l(r) = l(l+1)/r^2$, we have $\delta_l = 0$. Show that normalization according to

$$\int_0^\infty v_k v_{k'}\, r^2\, \mathrm{d}r = \delta(k'-k) \qquad [46.2]$$

is equivalent to the requirement

$$C = \sqrt{2/\pi}. \qquad [46.3]$$

For this purpose use [46.2] in the form

$$\lim_{R\to\infty}\int_0^R r^2\,\mathrm{d}r\int_{k-(\Delta k/2)}^{k+(\Delta k/2)} v_k v_{k'}\,\mathrm{d}k' = 1 \qquad [46.4]$$

and use the relation

$$\int_0^R v_k v_{k'} r^2\,\mathrm{d}r = \frac{1}{k'^2 - k^2}\left\{(rv_{k'})\frac{\mathrm{d}}{\mathrm{d}r}(rv_k) - (rv_k)\frac{\mathrm{d}}{\mathrm{d}r}(rv_{k'})\right\}_{r=R}, \qquad [46.5]$$

which follows from the continuity equation. The limit $R\to\infty$ must be taken after the integration over k'.

2. Show that this result is also valid in the case of the Coulomb potential if a correction of the order of $\log r$ is added to the phase in [46.1].

3. For the wave function for the Coulomb potential in parabolic coordinates, calculate the diagonal element of the coordinate $z = \frac{1}{2}(\lambda_1 - \lambda_2)$, Eq. [18.66] (expressed in terms of the unit of length a_0). For doing this, the generating function for the Laguerre polynomials $L^m_{m+n_l}$ can be used, as can the normalization integral that was already calculated in Section 17 (see Eq. [17.42]).

47. THE SYMMETRICAL TOP

Discuss the eigenvalue problem for the symmetrical top which has moments of inertia $A = B$ and C. The Hamiltonian function of classical mechanics, expressed in terms of the Euler angles ϑ, φ, χ and their canonically conjugate momenta p_ϑ, p_φ, p_χ, is

$$H = \frac{p_\vartheta^2}{2A} + \frac{(p_\varphi - \cos\vartheta\, p_\chi)^2}{2A \sin^2\vartheta} + \frac{p_\chi^2}{2C}.$$

This leads to the wave equation

$$\frac{1}{2A} \frac{1}{\sin\vartheta} \frac{\partial}{\partial\vartheta}\left(\sin\vartheta \frac{\partial\psi}{\partial\vartheta}\right) + \frac{1}{2A} \frac{1}{\sin^2\vartheta}\left(\frac{\partial}{\partial\varphi} - \cos\vartheta \frac{\partial}{\partial\chi}\right)^2 \psi$$

$$+ \frac{1}{2C} \frac{\partial^2\psi}{\partial\chi^2} = -\frac{E}{\hbar^2}\,\psi.$$

A trial solution

$$\psi = \theta(\vartheta) \exp[i(m\chi - m'\varphi)]$$

leads to the differential equation

$$\sin\vartheta \frac{d}{d\vartheta}\left(\sin\vartheta \frac{d\theta}{d\vartheta}\right)$$

$$- (m'^2 + m^2 + 2\cos\vartheta\, mm' - \lambda\sin^2\vartheta)\theta = 0, \qquad [47.1]$$

where

$$E = \frac{\hbar^2}{2}\left\{\frac{\lambda}{A} + m^2\left(\frac{1}{C} - \frac{1}{A}\right)\right\}. \qquad [47.2]$$

Substituting $x = (1 - \cos\vartheta)/2$, $1 - x = (1 + \cos\vartheta)/2$, we obtain

$$x(1-x)\frac{d}{dx}\left\{x(1-x)\frac{d\theta}{dx}\right\}$$

$$+ \left\{-\lambda x^2 - \frac{1}{4}(m+m')^2 + (\lambda + mm')x\right\}\theta = 0 \qquad [47.3]$$

from [47.1]. With

$$\theta_{m,m'} = x^{(m+m')/2}(1-x)^{(m-m')/2}\cdot f_{m,m'}, \qquad [47.4]$$

the equation for f is the hypergeometric differential equation

$$x(1-x)f'' + \{\gamma - (\alpha + \beta + 1)x\}f' - \alpha\beta f = 0, \qquad [47.5]$$

with

$$\left.\begin{array}{l} \alpha + \beta = 2m + 1 \\ \alpha\beta = -\lambda + m(m+1) \\ \gamma = m + m' + 1 \end{array}\right\}. \qquad [47.6]$$

If γ is neither zero nor a negative integer, then an integral of [47.5] which is regular at $x = 0$ is given by the series

$$F(\alpha, \beta, \gamma, x) = 1 + \frac{\alpha}{1}\frac{\beta}{\gamma}x + \frac{\alpha(\alpha+1)}{1\cdot 2}\cdot\frac{\beta(\beta+1)}{\gamma(\gamma+1)}x^2 + \cdots.$$

Here, we allow m and m' to have integral and half-integral values, of both signs, but such that $m \pm m'$ is integral.

Show that a solution valid for integral values of γ of both signs is given by the integral

$$\overline{F}(\alpha, \beta, \gamma, x)$$

$$= \frac{\exp[i\pi\alpha]}{2\pi i}\cdot\frac{\Gamma(1-\alpha)}{\Gamma(\gamma-\alpha)}\oint_C t^{\alpha-1}(1-t)^{\gamma-\alpha-1}(1-tx)^{-\beta}dt.$$

The path C encircles the points 0 and 1 in the positive sense, but the point $1/x$ is outside of C.

We have

$$F = \Gamma(\gamma)\overline{F}(\alpha, \beta, \gamma, x) \qquad \text{for } \gamma = 1, 2, \ldots.$$

The eigenvalues of λ come from the requirement that $\overline{F}$ be a polynomial and [47.4] be finite, which means that α is a negative integer or zero. (Interchanging α and β does not yield a new solution.) Show that this is the case if and only if

$$\lambda = j(j+1) \qquad [47.7]$$

(j is positive by definition), and

$$j \pm m \text{ and } j \pm m' \text{ are integral and nonnegative.} \quad [47.8]$$

We then have

$$\alpha = -j + m, \quad \beta = j + m + 1, \quad \gamma = m + m' + 1. \quad [47.9]$$

Show, by means of the two substitutions $t = 1/s$ and $t = (1/x)\big((x-s)/(1-s)\big)$ into the integral representation, that the function $\theta_{m,m'}$ in [47.4] with $f_{m,m'} = \overline{F}_{m,m'}$ satisfies

$$\theta_{-m,-m'} = (-1)^{m+m'} \cdot \frac{(j+m')! \, (j+m)!}{(j-m')! \, (j-m)!} \cdot \theta_{m,m'}, \qquad [47.10]$$

and derive

$$\overline{F} = \frac{(j-m)!}{(j+m)! \, (j+m')!} (-1)^{m+m'} \left(\frac{d}{dx}\right)^{j+m} \{x^{j-m'}(1-x)^{j+m'}\}$$

$$= \frac{1}{(j+m')!} x^{-m'-m}(1-x)^{m'-m}$$

$$\times \left(\frac{d}{dx}\right)^{j-m} \{x^{j+m'}(1-x)^{j-m'}\}. \qquad [47.11]$$

Moreover, show that

$$(1-x)^{\gamma-\alpha-\beta} F(\gamma-\beta, \gamma-\alpha, \gamma, x) = F(\alpha, \beta, \gamma, x)$$

implies

$$\theta_{m,m'} = \theta_{m',m}. \qquad [47.12]$$

Finally, calculate the normalization integral

$$N = \int_0^1 \{\theta_{m,m'}(x)\}^2 \, dx = \int_0^1 x^{m+m'}(1-x)^{m-m'} (\overline{F})^2 \, dx. \qquad [47.13]$$

In this expression, the $(\overline{F})^2$ can, for example, be replaced by the product of the two expressions for $\overline{F}$ given in [47.11]; then a partial integration can be performed, and the Euler integral of the first kind,

$$\int_0^1 x^{p-1}(1-x)^{q-1}\,\mathrm{d}x = \frac{\Gamma(p)\Gamma(q)}{\Gamma(p+q)}; \qquad p>0,\ q>0 \qquad [47.14]$$

can be used.

Bibliography

General

H. A. KRAMERS, *Quantum Mechanics* (North Holland, Amsterdam, 1957).

P. A. M. DIRAC, *The Principles of Quantum Mechanics*, 4th ed. (Oxford University Press, London, 1958).

E. C. KEMBLE, *The Fundamental Principles of Quantum Mechanics* (McGraw-Hill, New York, 1937); corrected republication by Dover Publications, Inc. (New York, 1958).

D. BOHM, *Quantum Theory* (Prentice-Hall, New York, 1951).

L. I. SCHIFF, *Quantum Mechanics*, 2nd ed. (McGraw-Hill, New York, 1955).

D. I. BLOCHINZEW, *Grundlagen der Quantenmechanik* (Deutscher Verlag der Wissenschaften, Berlin, 1953).

A. SOMMERFELD, *Atombau und Spektrallinien*, vol. 2 (Wellenmechanischer Erganzungsband), 2nd ed. (Vieweg, Braunschweig, 1944); reprinted by Frederick Ungar Publishing Co. (New York, 1953).

L. D. LANDAU and E. M. LIFSHITZ, *Quantum Mechanics, Nonrelativistic Theory* (Addison-Wesley, Reading, Mass., 1958).

A. MESSIAH, *Quantum Mechanics* (North Holland, Amsterdam, 1961), vols. 1 and 2.

Special

L. DE BROGLIE, *La Mecanique Ondulatoire* (Gauthier-Villars, Paris, 1928).

W. PAULI, "Die Allgemeinen Prinzipien der Wellenmechanik," article in *Encyclopedia of Physics*, Vol. 5, Part 1 (Springer, Berlin, 1958). This article is not introductory.

H. A. BETHE and E. E. SALPETER, "Quantum Mechanics of One- and Two-Electron Systems," article in *Encyclopedia of Physics*, Vol. 35 (Springer, Berlin, 1957).

E. Schrödinger, *Abhandlungen zur Wellenmechanik* (Leipzig, 1928).

M. Born and P. Jordan, *Elementare Quantenmechanik* (Springer, Berlin, 1930). This treatment is purely algebraic.

J. von Neumann, *Mathematical Foundations of Quantum Mechanics* (Princeton University Press, Princeton, 1955).

G. Ludwig, *Die Grundlagen der Quantenmechanik* (Springer, Berlin, 1954).

J. M. Jauch, *Foundations of Quantum Mechanics* (Addison-Wesley, Reading, Mass., 1968).

Discussions of Foundations and Epistemological Aspects

W. Heisenberg, *The Physical Principles of the Quantum Theory* (University of Chicago Press, Chicago, 1930); reprinted by Dover Publications, Inc. (New York, 1949).

N. Bohr, *Atomic Theory and the Description of Nature* (Cambridge University Press, London, 1934).

N. Bohr, "Discussion with Einstein on Epistemological Problems in Atomic Physics," article in *Albert Einstein: Philosopher-Scientist,* edited by P. A. Schilpp (Tudor, New York, 1951).

H. Reichenbach, *Philosophic Foundations of Quantum Mechanics* (University of California Press, Berkeley, 1944).

Dialectica, Vol. 2, number 3/4 (Neuchâtel, Switzerland, 1948).

W. Heisenberg, "The Development of the Interpretation of the Quantum Theory," article in *Niels Bohr and the Development of Physics,* edited by W. Pauli (McGraw-Hill, New York, 1955).

Institut International de Physique Solvay, Cinquième Conseil de Physique: *Electrons et Photons* (Paris, 1928).

Contributions by A. Einstein, E. Schrödinger, W. Pauli, and others in *Louis de Broglie physicien et penseur* (Albin Michel, Paris, 1953).

Appendix. Comments by the Editor

[A-1] (p. 12). The problem raised in this paragraph constitutes one of the most fascinating aspects of the history of quantum theory. It is the question of whether the probabilistic nature of quantum theory is compatible with the requirement that it give a complete description of nature. This problem was heatedly discussed at the Fifth Solvay congress in 1927 from which the "Copenhagen interpretation" emerged victorious. While the latter is a credo in this compatibility, the opposite view defended by de Broglie, Einstein, Schrödinger, and later by Bohm is a conviction in the deterministic foundation of a complete theory.

The "intuitive pictures" to which Pauli refers in this paragraph make allusion to the "guiding wave" of de Broglie, which is a special case of the "hidden variables" of the "deterministic scheme." The problem is most clearly exposed in Pauli's article "Remarques sur le problème des paramètres cachés dans la mécanique quantique et sur la théorie de l'onde pilote" in the volume dedicated to de Broglie's sixtieth birthday, *Louis de Broglie physicien et penseur* (Albin Michel, Paris, 1953). He writes (p. 35) "J'eus le plaisir de discuter cette théorie au congrès Solvay 1927 avec L. de Broglie. Peu après, de Broglie l'abandonna en faveur de l'interprétation complémentaire de la mécanique quantique de Heisenberg et Bohr; les raisons en sont exposées en détail dans son 'Introduction à la mécanique ondulatoire' (1929)." Later, on p. 37, Pauli proceeds to discuss from a general point of view "la tentative de compléter la mécanique quantique de façon à en faire un schéma déterministe à l'aide de paramètres cachés; la théorie de l'onde pilote n'en est qu'un example spécial." And he concludes (p. 42) "Ce sont ces raisons physiques, qui n'ont rien à faire avec des préjugés philosophiques quant à l'interprétation et à la justification de théories physiques en général, qui me font juger que l'interprétation de la mécanique quantique basée sur l'idée de la complémentarité est la seule admissible. Loin de considérer comme définitif l'état actuel de la mécanique quantique dans le domaine relativiste, je crois pourtant que le développement de cette théorie ne fera que nous éloigner davantage de la possibilité d'une interprétation causale et déterministe."

As to Einstein's belief mentioned in this paragraph it is most clearly expressed in his article "Einleitende Bemerkungen ueber Grundbegriffe" in the same volume. He writes (p. 6) *"Es gibt so etwas wie den 'realen Zustand'* eines physikalischen Systems, was unabhängig von jeder Beobachtung oder Messung objectiv existiert und mit den Ausdrucksmitteln der Physik im Prinzip beschrieben werden kann." And p. 8, "Man fühlt sich daher gezwungen, die Beschreibung eines Systems durch eine Ψ-Funktion als eine unvollständige Beschreibung des realen Zustandes zu betrachten."

For a more recent discussion of the problem of hidden variables see J. S. Bell, *Rev. Mod. Phys.* **38**, 447 (1966). For a very recent experimental test, see S. J. Freedman and J. F. Clauser, *Phys. Rev. Letters* **28**, 938 (1972).

[A-2] (p. 28). The $\delta(x)$ and all its derivatives are tempered distributions.

A distribution is defined as a linear continuous functional $F[\varphi]$ of indefinitely differentiable functions $\varphi(x)$ with bounded support. This means

$$F[\lambda_1\varphi_1 + \lambda_2\varphi_2] = \lambda_1 F[\varphi_1] + \lambda_2 F[\varphi_2],$$

$$\lim_{j\to\infty} F[\varphi_j] = F[\varphi],$$

for any sequence $\varphi_1, \varphi_2, \ldots$ such that $\lim_{j\to\infty} \varphi_j = \varphi$, and the nonzero values of $\varphi(x)$ are all contained in a finite domain of x.

A tempered distribution is defined with functions $\psi(x)$ all derivatives $\psi^{(m)}$ of which tend asymptotically strongly to zero,

$$\lim_{|x|\to\infty} |x|^l \psi^{(m)}(x) = 0$$

for any nonnegative l and m.

A locally integrable function $f(x)$ which increases sufficiently slowly at infinity (i.e., for which there exist positive numbers A and α such that $|f(x)| \leqslant Ax^\alpha$ when $|x| \to \infty$) defines a tempered distribution

$$f[\psi] = \int f(x)\,\psi(x)\,\mathrm{d}x\,.$$

Two locally integrable functions f_1 and f_2 define the same distribution if $f_1 - f_2 = 0$ except on a set of measure zero. Therefore all representations of the Dirac function $\delta(x)$ of which examples are given on p. 29 define the same distribution

$$\delta[\psi] = \int \delta(x)\,\psi(x)\,\mathrm{d}x = \psi(0)\,.$$

If a locally integrable function is differentiable, the derivative of its distribution is defined by

$$f'[\psi] = \int f'(x)\,\psi(x)\,\mathrm{d}x = -\int f(x)\,\psi'(x)\,\mathrm{d}x = -f[\psi']\,.$$

In quantum mechanics the scalar product of an eigenfunction f of an observable with continuous spectrum with a general wave function $\psi(x)$, $\int f(x)\psi(x)\,dx$, defines a tempered distribution $f[\psi]$ because any wave function $\psi(x)$ is square integrable.

For a concise exposition of the properties of distributions see, e.g., A. Messiah, *Quantum Mechanics* (North Holland, Amsterdam, 1961), vol. 1, appendix A.

[A-3] (pp. 156, 175). Pauli's role in the history of the spin both of the electron and of nuclei is most significant. As early as 1924 he proposed the hypothesis of a "resulting angular momentum" of nuclei. Indeed, in *Naturwiss.* 12, 741 (1924) he discusses the "satellites of some spectral lines" in terms of what today is called the hyperfine interaction; he writes "Diese Energieunterschiede werden dabei als vom zusammengesetzten Bau des Kernes herrührend aufgefasst und es wird angenommen, dass der Kern im allgemeinen ein nicht verschwindendes resultierendes Impulsmoment besitzt."

It is remarkable that at the same time Pauli was not ready to accept the idea of an angular momentum of the electron, although in *Z. Physik* 31, 373 (1925) (submitted: 2 December 1924) he drew attention to the "strange two-valuedness" of the quantum states of the electron. This conclusion was based both on the "doublet structure of the alkali spectra" (spin-orbit splitting) and on the "violation of Larmor's theorem" (anomalous Zeeman effect). He writes in the quoted paper "Die Dublettstruktur der Alkalispektren, sowie die Durchbrechung des Larmortheorems kommt gemäss diesem Standpunkt durch eine eigentümliche, klassisch nicht beschreibbare Art von Zweideutigkeit der quantentheoretischen Eigenschaften des Leuchtelektrons zustande." Pauli explained the reason for his hesitation to accept the idea of the electron spin in his 1946 Nobel Prize Lecture (reprinted in *Collected Scientific Papers by Wolfgang Pauli*, Interscience, New York, 1964, vol. 2, p. 1080) as follows: "Although at first I strongly doubted the correctness of this idea because of its classical mechanical character, I was finally converted to it by Thomas' calculation [L. H. Thomas, *Nature* **117**, 514 (1926) and *Phil. Mag.* **3**, 1 (1927). Compare also J. Frenkel, *Z. Phys.* **37**, 243 (1926)] on the magnitude of doublet splitting ... ."

It was indeed because of the discrepancy by this Thomas factor of two that Pauli did not believe in Kronig's calculation of the spin-orbit splitting (unpublished, beginning 1925). More details on this subtle episode of the history of science are found in the articles by R. Kronig and by B. L. van der Waerden in *Theoretical Physics in the Twentieth Century. A Memorial Volume to Wolfgang Pauli* (Interscience, New York, 1960).

With regard to the history of the nuclear spin an amusing account can be found in the article by S. A. Goudsmit in *Physics Today* **14**, No. 6, p. 18 (1961); he writes "For a number of years, whenever I met Pauli, he would remark cryptically that he 'could afford not to be quoted.' It was only in the late thirties that I found out to what he referred."

[A-4] (p. 174). This is true for spin-independent forces only. In this case the scattering does not give rise to spin flip so that identical particles with antiparallel spins remain distinguishable in the collision.

[A-5] (p. 178). This problem is treated in great detail by M. Born in *Dan. Mat. Fys. Medd.* **30**, No. 2 (1955). This paper was dedicated to Niels Bohr on the occasion of his seventieth birthday.

Index

Airy functions, 128
Alkali atom, 97
α particles, 174–175
Angular momentum, 92
  commutation relation of, 152, 153
  components of, in polar
    coordinates, 184
  eigenvalue equation for, 153
  matrix elements of, 154–155, 184
  total, 159
Anharmonic oscillator, isotropic, 74
Anomalous Zeeman splitting, 156
Antisymmetric particles
  (fermions), 166
Antisymmetric state, 170
Asymptotic formula,
  for confluent hypergeometric
    function, 87, 95, 99, 108
  for hydrogen atom, $(\varepsilon > 0)$, 100
  for partial wave, 117, 123, 185
Asymptotic series, 86, 87, 133
Asymptotic solution,
  for Coulomb potential, 109–110
  for particles in homogenous
    field, 130
  for plane harmonic oscillator, 88
  of scattering problem, 108–110,
    117–118
Auger effect, 149
Average over phases, 22
Average values, 5, 6, 46

Balmer formula, 97
Band spectra of molecules, 175

Bessel functions, 114
Bessel's inequality, 32
Bohm, D., 193
Bohr, N., 14
Bohr quantum condition, 136
Bohr radius, 93
Born, M., 69, 153, 197
Born approximation, 105, 120–123
Bose statistics, 166, 175
Bosons, 166, 167, 174
Boundary conditions, 24, 177
  radiation, 120
Bound states, 179
Brillouin, L., 132
Burhop, E. H. S., 149

Canonical equation of motion,
  classical, 38, 46
Cauchy and Riemann, method of
  steepest descent of, 129
Central force field (potential),
  88, 90, 116–120
Centrifugal force, 92
Classical canonical equations of
  motion, 46
Classical mechanics, 3, 13, 20–21,
  22, 38, 43, 46, 133
Classical quantities, 3, 39
Classical turning points, 134, 136
Coherence property of light, 18–19
Collision of two identical
  particles, 173–175
Collision processes, 107–125

Commutation relations, 44, 56, 64, 66, 69
  for spin, 158
  of angular momentum, 152
Commutator, 44
Complementary quantities, 14
Completeness of Hermitian polynomials, 60–62, 183
Completeness relation, 32, 33, 34, 67, 139, 177
Complete orthonormal set of functions, 31, 63, 67, 138, 139
Confluent hypergeometric function, 81–85
Conservation,
  of energy, 14, 47
  of momentum, 14
Continuity equation, 26, 48
Continuous spectrum,
  of hydrogen atom, 98–100, 105–106
  matrix method for, 142
  in perturbation theory, 149–151
Convergence in the mean, 32, 33
Coulomb integral, 171
Coulomb interaction, 54, 169
Coulomb potential (field), 92, 97, 100, 102, 109–110, 111, 165, 171, 173, 185, 86
Courant, R. and Hilbert, D., 129
Cross section,
  for collisions of two identical particles, 174
  differential scattering, 110, 118
  for low-energy particles, 125
  total, 119
Current density, 25, 48, 111
Cylinder functions, 114, 128

de Broglie, L., 2, 193
Degeneracy,
  degree of, 27, 97
  for hydrogen atom, 96–97
  for isotropic harmonic oscillator, 72–74
  in perturbation theory, 146–147
δ-function, 33, 194, 195
Diaphragm, 12, 18, 21

Differential cross section, 118, 174
Differential equation,
  for confluent hypergeometric function, 81
  for cylinder functions, 114
  for eikonal, 132
  for general hypergeometric function, 82, 187
  for Hermite polynomials, 56
  for Laguerre polynomials, 76
  radial, 92, 117
  Riccati, 133
  for spherical harmonics, 89, 90, 181
Differential operator, 3, 153
Diffraction, theory of, 126, 128
Diffraction experiment, 18
Diffraction pattern, 12
Dirac, P. A. M., 147, 152, 153, 156
Dirac δ-function, 28, 142, 194, 195
Dirac equation, 156
Discrete energy spectrum of hydrogen atom, 95–97, 104–105
Distribution,
  Gaussian, 8
  tempered, 194, 195
Doppler effect, 16–17

Ehrenfest, P., 161
Eigenfunctions, 26, 142
  of periodic potential, 180
  spin, 169
Eigenstates, 27
Eigenvalue, 27
Eigenvalue problem, 25, 55, 141–142
  of particle in uniform field 126–132
  of symmetrical top, 186–189
Eikonal equation, 132
Einstein, A., 12, 193, 194
Electron, 93, 149, 156, 157, 158, 159, 165, 169, 170, 171, 172, 174–175, 176
  spin of, 156, 157
Energy, conservation of, 14, 47
Energy eigenvalues, 66, 93, 141, 142, 144

Energy shell, 150
Equation of continuity, 26
Equation of motion, canonical, 38, 46
Euler angles, 161, 186
Euler integral (of the first kind), 189
Euler integral representation (of Γ function), 81
Exchange integral, 172
Exchange operator, 165
Exclusion principle, 167–169
Expectation value, 11, 40, 43

Fermi, "golden rule" of, 151
Fermions, 166, 167, 173–175
Fermi statistics, 166
Field,
  of force, 53, 55, 88
  uniform, 126
Fourier integral (transformation), 5, 8, 27
Frequency spectrum, 8
Fresnel integral, 35
Fundamental solution. *See* Initial value problem

Γ function, 80, 100
  integral representation of, 81
Gauge group, 49
Gauge invariance, 48–49
Gaussian distribution, 8, 178
Generating function,
  of Hermite polynomials, 58
  of Laguerre polynomials, 78
"Golden rule," 151
Gordon, W., 111
Goudsmit, S., 197
Green's formula, 26, 42
Ground state,
  of helium atom, 171
  of hydrogen atom, 97
Group theory, 164
Group velocity, 2

Hamiltonian function (classical), 38, 39, 186

Hamiltonian matrix, 66, 69, 143
Hamiltonian operator, 39, 42
  for two electrons in a magnetic field, 159
  for uncoupled particles, 53
Hankel functions, 114
Hankel relation (of Γ function), 81
Harmonic oscillator,
  anisotropic, 73
  eigenvalues of linear, 57, 66
  fundamental solution for linear, 182–183
  isotropic (in plane), 72, 73, 87, 88
  linear, 55, 137, 142
  linear, with additional potential, 67–68
  matrices for linear, 65, 66
  matrix method for linear, 69–72
  Schrödinger equation for isotropic, 72, 74
  Schrödinger equation for linear, 56
  selection rule for linear, 70
Harmonic polynomial, 180
Harmonics,
  spherical, 89–91, 154, 180–182
  tesseral, 91
  zonal, 91
Heat conduction, 4, 37
Heisenberg, W., 69, 172
Heisenberg uncertainty relation, 14
Helium,
  ortho, 171
  para, 171
Helium atom, 169–172
Helium spectrum, 169, 172
Hermite polynomials, 56, 183
Hermitian matrix, 63
Hermitian operator, 40, 41, 42–43
Hermiticity (condition), 42, 63, 138, 147
Hertz dipole, 120
"Hidden variables," 193–194
Hilbert space, 66, 67
Homogeneous polynominal, 91

Hydrogen atom,
  angular momentum quantum
    number of, 96
  continuous energy spectrum of,
    98–100, 105–106
  degeneracy of, 96–97
  discrete energy spectrum of,
    95–97, 104–105
  energy eigenvalue of, 96, 104, 105
  principal quantum number of,
    96, 97
  radial quantum number of, 95
  separation of, in parabolic
    coordinates of, 102–106
  separation of, in spherical
    coordinates, 88–89
Hyperfine interaction, 196
Hypergeometric function,
  confluent, 81–85, 94–96, 107
  general, 82
  integral representation of, 82

Incident plane wave, 107, 111, 120
Initial value problem, 33–37, 177–
    178, 182–183
Integral representation,
  of confluent hypergeometric
    function, 83, 85, 98
  of cylinder functions, 114
  Euler, of Γ function, 81
  of Laguerre polynomials, 77
  of spherical harmonics, 184
Interaction,
  Coulomb, 54, 169, 171
  between particles, 51
  spin-orbit, 169
Interaction potential, 60
Interference effects (phenomena),
    12, 14
Interference of probabilities, 20,
    178
Interference term, 174–175, 178

Jeffreys, H., 132
Jordan, P., 69, 153

Kemble, E. C., 137

Klein, F., 160
Kramers, H. A., 132
Kronig, R., 197
Kronig-Penney potential, 180

Laguerre polynomials, 76, 84, 96,
    186
Laplacian,
  in orthogonal curvilinear
    coordinates, 102
  in parabolic coordinates, 103
  in spherical coordinates, 88
Legendre polynomials, 91, 92, 116,
    122
Light, coherence property of, 18–19
Light quanta, 1, 15
Linear harmonic oscillator, 55
Linear operator, 41
Linear superposition of
    (stationary) states, 24, 27, 34
Lorentz group, 164

Magnetic field, 38, 42, 46, 47, 74,
    97, 159, 165
Matrix,
  Hermitian, 63
  unitary, 140, 160
Matrix element, 63, 138
  of angular momentum, 154
  of harmonic oscillator, 64
Matrix mechanics
Matrix method,
  for continuous spectra, 142
  of harmonic oscillator, 69–72
Matrix representation, 68, 69
Measurement,
  momentum, 16–17
  position, 14–15
Messiah, A., 196
Method of images, 177
Microscope, 14–15
Mixture, 21–22
Moment of inertia, 175, 186
Momentum conservation, 14
Momentum measurement, 16–17
Mott, N. F., 173

Neutron, 168, 176
Nonrelativistic wave equation, 4, 37
Normalization, 5
in the continuum, 27–31
of Hermite polynomials, 58
for plane harmonic oscillator, 79
of spherical harmonics, 91, 182
for spinning electron, 159
for symmetrical top, 188
Nuclear spin, 175–176, 196–197
Nucleus (atomic), 93, 176

Observable quantities, 43
Operator,
differential, 3, 153
exchange, 165
Hamiltonian, 39, 42
Hermitian, 40, 41, 42–43
linear, 41
transformation, 139
Orthogonality relation, 25, 26, 28, 58
for spin, 157
Ortho helium, 171
Orthonormal functions, 31
Ortho states, 171

Parabolic coordinates, 101, 186
Para helium, 171
Para states, 171
Parseval, formula of, 5
Partial waves, method of, 117–118, 185–186
Particle,
in box, 23–27
in magnetic field, 38, 39
Particle current density, 111
Pauli, W., 156, 164, 166, 168, 176, 193–194, 196–197
Pauli spin matrices, 158
Periodic potential, 180
Perturbation theory, 67, 68, 121, 142
in matrix representation, 143–147
time-dependent, 147–151
Phase correction, logarithmic, 110

Phase integral, 136
Phase shift, 119–120
Phase velocity, 2
Polar coordinates,
plane, 73
spherical, 88
Position measurement, 14–15
Potential, 42, 54, 67
central, 88, 90, 116–120
Coulomb, 92, 97, 100, 102, 109–110, 111, 173
periodic, 180
range of, 122
vector, 38
Potential barrier, 137
Potential well, 179
Probabilistic nature of quantum theory, 193
Probability, reduction of, 19, 21
Probability current density, 25, 48
Probability density, 11, 19–20, 24, 52
Probability distribution, 43, 178
Proton, 168, 175, 176
Pure case, 21, 22

Quantum number,
angular momentum, 96
principal, 96
radial, 95

Radial differential equation, 92, 117
Range of potential, 122
Recoil momentum, 18
Recursion formula,
for Hermite polynomials, 60
for Laguerre polynomials, 82
for spherical harmonics, 181
Reduced mass, 93, 173
Relativistic particle mechanics, 1
Relativistic spin theory, 164
Relativistic wave equation (scalar), 3
Representation, matrix, 68, 69
Rotation group, 153
Rotation in space, 160–164
Rotator ("dumbbell"), 175

Russel-Saunders coupling, 169
Rutherford scattering formula, 111

Saddle point method, 128
Scattered wave, 107, 111, 120, 121
Scattering center, 118
Scattering cross section, 110–111,
    118–119
Scattering formula, Rutherford, 111
Scattering of low-energy particles,
    123–125
Schrödinger E., 39, 55, 79, 193
Schrödinger equation, 39, 55, 68,
    88. See also Wave equation
Secular equation, 147
Selection rule for harmonic
    oscillator, 70
Separable equation, 72, 73
Sexl, Th., 111
Similarity transformation, 141
Singlet, 170, 171
Singlet scattering amplitudes, 174
Sommerfeld, A., 101, 105, 160
Spectral analysis, 8
Spherical harmonics, 89–91, 154,
    175, 180–182
Spherical wave,
  incoming, 112
  outgoing, 108, 112
Spin,
  history of, 197
  of electron, 156, 157
Spin density, 162
Spin eigenfunctions, 169
Spin matrices, 158
Spin operators, 156
Spinor, 161
Spin-orbit interaction, 169, 172
Spin variable, 158
Standing waves, 23, 55
Stark effect (first order), 105
State, 11, 21
  antisymmetric, 170
  bound, 179
  ground, 171
  ortho, 171
  para, 171

singlet, 170
stationary, 23, 24
symmetric, 170
triplet, 170, 172
Stationary state, 23, 24
Statistics,
  Bose, 166
  classical and quantum, 19–22
  Fermi, 166
  of nuclear spins, 175–176
  of rotational states, 175
Superposition principle, 11, 40
Symmetrical top, 186–189
Symmetric particles (bosons), 166
Symmetric state, 170
Symmetry classes, 165–167, 170
Symmetry of Hamiltonian
    operator, 165

ϑ function, 177
Time-dependent forces, 46, 148
Time-dependent perturbation
    theory, 147–151
Total cross section, 119, 125
Transformation,
  principal axis, 146
  similarity, 141
  unitary, 140
Transformation operator, 139
Transformation theory, 138–143
Transition probability, 150–151
Translational key between classical
    and wave mechanics, 3, 38
Transmission coefficient, 179
Triplet, 170, 171
Triplet scattering amplitudes, 174
Tunnel effect, 137, 179

Uhlenbeck, G. E. and Goudsmit, S., 156
Uncertainty,
  of momentum, 15
  in position, 17
Uncertainty relation, 8, 14, 19, 20
Unitary matrix, 140, 160
Unitary transformation, 140, 141

Valence electron, 97

van der Waerden, B. L., 164, 197
Variance of Gaussian distribution, 9, 11
Vector, in Hilbert space, 67
Vector potential, 38
Velocity,
    group, 2
    particle, 2
    phase, 2
Virial theorem, 49

Wave,
    incident plane, 107, 111, 120
    scattered, 107, 111, 120, 121
    standing, 23
Wave equation,
    for central potential, 88, 116
    force-free, 112–114
    for harmonic oscillator in plane, 72
    for linear harmonic oscillator, 56
    nonrelativistic, 4, 37
    relativistic scalar, 3
    for symmetrical top, 186
    time-dependent, 52, 147
    time-independent, 39, 55, 132
    with uniform field, 126
Wave function, 3, 34
    antisymmetric, 165, 171
    generalized, 51
    symmetric, 165
    two-component, 161
Wave kinematics, 14
Wave nature of atom, 19
Wave normal, 1
Wave number, 10
Wave packet, 3, 5, 11, 17, 178
    with Gaussian distribution, 8
    normalized, 5
    overlapping, 168
Wave train, 16, 17
Wave vector, 1
Wentzel, G., 132
Whittaker, E. T. and Watson, G. N., 86, 177
Wigner, E. P., 164
WKB method, 132–137

Zeeman splitting, anomalous, 156
Zero-point energy, 57, 72

A CATALOG OF SELECTED
DOVER BOOKS
IN SCIENCE AND MATHEMATICS

A CATALOG OF SELECTED

DOVER BOOKS

IN SCIENCE AND MATHEMATICS

Astronomy

BURNHAM'S CELESTIAL HANDBOOK, Robert Burnham, Jr. Thorough guide to the stars beyond our solar system. Exhaustive treatment. Alphabetical by constellation: Andromeda to Cetus in Vol. 1; Chamaeleon to Orion in Vol. 2; and Pavo to Vulpecula in Vol. 3. Hundreds of illustrations. Index in Vol. 3. 2,000pp. 6⅛ x 9¼.
23567-X, 23568-8, 23673-0 Three-vol. set

THE EXTRATERRESTRIAL LIFE DEBATE, 1750–1900, Michael J. Crowe. First detailed, scholarly study in English of the many ideas that developed from 1750 to 1900 regarding the existence of intelligent extraterrestrial life. Examines ideas of Kant, Herschel, Voltaire, Percival Lowell, many other scientists and thinkers. 16 illustrations. 704pp. 5⅜ x 8½. 40675-X

A HISTORY OF ASTRONOMY, A. Pannekoek. Well-balanced, carefully reasoned study covers such topics as Ptolemaic theory, work of Copernicus, Kepler, Newton, Eddington's work on stars, much more. Illustrated. References. 521pp. 5⅜ x 8½.
65994-1

AMATEUR ASTRONOMER'S HANDBOOK, J. B. Sidgwick. Timeless, comprehensive coverage of telescopes, mirrors, lenses, mountings, telescope drives, micrometers, spectroscopes, more. 189 illustrations. 576pp. 5⅝ x 8¼. (Available in U.S. only.)
24034-7

STARS AND RELATIVITY, Ya. B. Zel'dovich and I. D. Novikov. Vol. 1 of *Relativistic Astrophysics* by famed Russian scientists. General relativity, properties of matter under astrophysical conditions, stars, and stellar systems. Deep physical insights, clear presentation. 1971 edition. References. 544pp. 5⅝ x 8¼. 69424-0

Chemistry

CHEMICAL MAGIC, Leonard A. Ford. Second Edition, Revised by E. Winston Grundmeier. Over 100 unusual stunts demonstrating cold fire, dust explosions, much more. Text explains scientific principles and stresses safety precautions. 128pp. 5⅜ x 8½. 67628-5

THE DEVELOPMENT OF MODERN CHEMISTRY, Aaron J. Ihde. Authoritative history of chemistry from ancient Greek theory to 20th-century innovation. Covers major chemists and their discoveries. 209 illustrations. 14 tables. Bibliographies. Indices. Appendices. 851pp. 5⅜ x 8½. 64235-6

CATALYSIS IN CHEMISTRY AND ENZYMOLOGY, William P. Jencks. Exceptionally clear coverage of mechanisms for catalysis, forces in aqueous solution, carbonyl- and acyl-group reactions, practical kinetics, more. 864pp. 5⅜ x 8½.
65460-5

THE HISTORICAL BACKGROUND OF CHEMISTRY, Henry M. Leicester. Evolution of ideas, not individual biography. Concentrates on formulation of a coherent set of chemical laws. 260pp. 5⅜ x 8½. 61053-5

A SHORT HISTORY OF CHEMISTRY, J. R. Partington. Classic exposition explores origins of chemistry, alchemy, early medical chemistry, nature of atmosphere, theory of valency, laws and structure of atomic theory, much more. 428pp. 5⅜ x 8½. (Available in U.S. only.) 65977-1

GENERAL CHEMISTRY, Linus Pauling. Revised 3rd edition of classic first-year text by Nobel laureate. Atomic and molecular structure, quantum mechanics, statistical mechanics, thermodynamics correlated with descriptive chemistry. Problems. 992pp. 5⅜ x 8½. 65622-5

Engineering

DE RE METALLICA, Georgius Agricola. The famous Hoover translation of greatest treatise on technological chemistry, engineering, geology, mining of early modern times (1556). All 289 original woodcuts. 638pp. 6¾ x 11. 60006-8

FUNDAMENTALS OF ASTRODYNAMICS, Roger Bate et al. Modern approach developed by U.S. Air Force Academy. Designed as a first course. Problems, exercises. Numerous illustrations. 455pp. 5⅜ x 8½. 60061-0

DYNAMICS OF FLUIDS IN POROUS MEDIA, Jacob Bear. For advanced students of ground water hydrology, soil mechanics and physics, drainage and irrigation engineering and more. 335 illustrations. Exercises, with answers. 784pp. 6⅛ x 9¼. 65675-6

ANALYTICAL MECHANICS OF GEARS, Earle Buckingham. Indispensable reference for modern gear manufacture covers conjugate gear-tooth action, gear-tooth profiles of various gears, many other topics. 263 figures. 102 tables. 546pp. 5⅜ x 8½. 65712-4

MECHANICS, J. P. Den Hartog. A classic introductory text or refresher. Hundreds of applications and design problems illuminate fundamentals of trusses, loaded beams and cables, etc. 334 answered problems. 462pp. 5⅜ x 8½. 60754-2

MECHANICAL VIBRATIONS, J. P. Den Hartog. Classic textbook offers lucid explanations and illustrative models, applying theories of vibrations to a variety of practical industrial engineering problems. Numerous figures. 233 problems, solutions. Appendix. Index. Preface. 436pp. 5⅜ x 8½. 64785-4

STRENGTH OF MATERIALS, J. P. Den Hartog. Full, clear treatment of basic material (tension, torsion, bending, etc.) plus advanced material on engineering methods, applications. 350 answered problems. 323pp. 5⅜ x 8½. 60755-0

A HISTORY OF MECHANICS, René Dugas. Monumental study of mechanical principles from antiquity to quantum mechanics. Contributions of ancient Greeks, Galileo, Leonardo, Kepler, Lagrange, many others. 671pp. 5⅜ x 8½. 65632-2

METAL FATIGUE, N. E. Frost, K. J. Marsh, and L. P. Pook. Definitive, clearly written, and well-illustrated volume addresses all aspects of the subject, from the historical development of understanding metal fatigue to vital concepts of the cyclic stress that causes a crack to grow. Includes 7 appendixes. 544pp. 5⅜ x 8½.   40927-9

STATISTICAL MECHANICS: Principles and Applications, Terrell L. Hill. Standard text covers fundamentals of statistical mechanics, applications to fluctuation theory, imperfect gases, distribution functions, more. 448pp. 5⅜ x 8½.   65390-0

THE VARIATIONAL PRINCIPLES OF MECHANICS, Cornelius Lanczos. Graduate level coverage of calculus of variations, equations of motion, relativistic mechanics, more. First inexpensive paperbound edition of classic treatise. Index. Bibliography. 418pp. 5⅜ x 8½.   65067-7

THE VARIOUS AND INGENIOUS MACHINES OF AGOSTINO RAMELLI: A Classic Sixteenth-Century Illustrated Treatise on Technology, Agostino Ramelli. One of the most widely known and copied works on machinery in the 16th century. 194 detailed plates of water pumps, grain mills, cranes, more. 608pp. 9 x 12.
          28180-9

ORDINARY DIFFERENTIAL EQUATIONS AND STABILITY THEORY: An Introduction, David A. Sánchez. Brief, modern treatment. Linear equation, stability theory for autonomous and nonautonomous systems, etc. 164pp. 5⅜ x 8¼.
          63828-6

ROTARY WING AERODYNAMICS, W. Z. Stepniewski. Clear, concise text covers aerodynamic phenomena of the rotor and offers guidelines for helicopter performance evaluation. Originally prepared for NASA. 537 figures. 640pp. 6⅛ x 9¼.
          64647-5

INTRODUCTION TO SPACE DYNAMICS, William Tyrrell Thomson. Comprehensive, classic introduction to space-flight engineering for advanced undergraduate and graduate students. Includes vector algebra, kinematics, transformation of coordinates. Bibliography. Index. 352pp. 5⅜ x 8½.   65113-4

HISTORY OF STRENGTH OF MATERIALS, Stephen P. Timoshenko. Excellent historical survey of the strength of materials with many references to the theories of elasticity and structure. 245 figures. 452pp. 5⅜ x 8½.   61187-6

ANALYTICAL FRACTURE MECHANICS, David J. Unger. Self-contained text supplements standard fracture mechanics texts by focusing on analytical methods for determining crack-tip stress and strain fields. 336pp. 6⅛ x 9¼.   41737-9

Mathematics

HANDBOOK OF MATHEMATICAL FUNCTIONS WITH FORMULAS, GRAPHS, AND MATHEMATICAL TABLES, edited by Milton Abramowitz and Irene A. Stegun. Vast compendium: 29 sets of tables, some to as high as 20 places. 1,046pp. 8 x 10½.   61272-4

FUNCTIONAL ANALYSIS (Second Corrected Edition), George Bachman and Lawrence Narici. Excellent treatment of subject geared toward students with background in linear algebra, advanced calculus, physics and engineering. Text covers introduction to inner-product spaces, normed, metric spaces, and topological spaces; complete orthonormal sets, the Hahn-Banach Theorem and its consequences, and many other related subjects. 1966 ed. 544pp. 6⅛ x 9¼. 40251-7

ASYMPTOTIC EXPANSIONS OF INTEGRALS, Norman Bleistein & Richard A. Handelsman. Best introduction to important field with applications in a variety of scientific disciplines. New preface. Problems. Diagrams. Tables. Bibliography. Index. 448pp. 5⅜ x 8½. 65082-0

FAMOUS PROBLEMS OF GEOMETRY AND HOW TO SOLVE THEM, Benjamin Bold. Squaring the circle, trisecting the angle, duplicating the cube: learn their history, why they are impossible to solve, then solve them yourself. 128pp. 5⅜ x 8½. 24297-8

VECTOR AND TENSOR ANALYSIS WITH APPLICATIONS, A. I. Borisenko and I. E. Tarapov. Concise introduction. Worked-out problems, solutions, exercises. 257pp. 5⅝ x 8¼. 63833-2

THE ABSOLUTE DIFFERENTIAL CALCULUS (CALCULUS OF TENSORS), Tullio Levi-Civita. Great 20th-century mathematician's classic work on material necessary for mathematical grasp of theory of relativity. 452pp. 5⅝ x 8¼. 63401-9

AN INTRODUCTION TO ORDINARY DIFFERENTIAL EQUATIONS, Earl A. Coddington. A thorough and systematic first course in elementary differential equations for undergraduates in mathematics and science, with many exercises and problems (with answers). Index. 304pp. 5⅜ x 8½. 65942-9

FOURIER SERIES AND ORTHOGONAL FUNCTIONS, Harry F. Davis. An incisive text combining theory and practical example to introduce Fourier series, orthogonal functions and applications of the Fourier method to boundary-value problems. 570 exercises. Answers and notes. 416pp. 5⅜ x 8½. 65973-9

COMPUTABILITY AND UNSOLVABILITY, Martin Davis. Classic graduate-level introduction to theory of computability, usually referred to as theory of recurrent functions. New preface and appendix. 288pp. 5⅜ x 8½. 61471-9

ASYMPTOTIC METHODS IN ANALYSIS, N. G. de Bruijn. An inexpensive, comprehensive guide to asymptotic methods–the pioneering work that teaches by explaining worked examples in detail. Index. 224pp. 5⅜ x 8½ 64221-6

ESSAYS ON THE THEORY OF NUMBERS, Richard Dedekind. Two classic essays by great German mathematician: on the theory of irrational numbers; and on transfinite numbers and properties of natural numbers. 115pp. 5⅜ x 8½. 21010-3

APPLIED COMPLEX VARIABLES, John W. Dettman. Step-by-step coverage of fundamentals of analytic function theory–plus lucid exposition of five important applications: Potential Theory; Ordinary Differential Equations; Fourier Transforms; Laplace Transforms; Asymptotic Expansions. 66 figures. Exercises at chapter ends. 512pp. 5⅜ x 8½.   64670-X

INTRODUCTION TO LINEAR ALGEBRA AND DIFFERENTIAL EQUA-TIONS, John W. Dettman. Excellent text covers complex numbers, determinants, orthonormal bases, Laplace transforms, much more. Exercises with solutions. Undergraduate level. 416pp. 5⅜ x 8½.   65191-6

MATHEMATICAL METHODS IN PHYSICS AND ENGINEERING, John W. Dettman. Algebraically based approach to vectors, mapping, diffraction, other topics in applied math. Also generalized functions, analytic function theory, more. Exercises. 448pp. 5⅜ x 8¼.   65649-7

CALCULUS OF VARIATIONS WITH APPLICATIONS, George M. Ewing. Applications-oriented introduction to variational theory develops insight and promotes understanding of specialized books, research papers. Suitable for advanced undergraduate/graduate students as primary, supplementary text. 352pp. 5⅜ x 8½.
                                                                      64856-7

COMPLEX VARIABLES, Francis J. Flanigan. Unusual approach, delaying complex algebra till harmonic functions have been analyzed from real variable viewpoint. Includes problems with answers. 364pp. 5⅜ x 8½.   61388-7

AN INTRODUCTION TO THE CALCULUS OF VARIATIONS, Charles Fox. Graduate-level text covers variations of an integral, isoperimetrical problems, least action, special relativity, approximations, more. References. 279pp. 5⅜ x 8½.
                                                                      65499-0

CATASTROPHE THEORY FOR SCIENTISTS AND ENGINEERS, Robert Gilmore. Advanced-level treatment describes mathematics of theory grounded in the work of Poincaré, R. Thom, other mathematicians. Also important applications to problems in mathematics, physics, chemistry and engineering. 1981 edition. References. 28 tables. 397 black-and-white illustrations. xvii + 666pp. 6⅛ x 9¼.
                                                                      67539-4

INTRODUCTION TO DIFFERENCE EQUATIONS, Samuel Goldberg. Exceptionally clear exposition of important discipline with applications to sociology, psychology, economics. Many illustrative examples; over 250 problems. 260pp. 5⅜ x 8½.
                                                                      65084-7

NUMERICAL METHODS FOR SCIENTISTS AND ENGINEERS, Richard Hamming. Classic text stresses frequency approach in coverage of algorithms, polynomial approximation, Fourier approximation, exponential approximation, other topics. Revised and enlarged 2nd edition. 721pp. 5⅜ x 8½.   65241-6

INTRODUCTION TO NUMERICAL ANALYSIS (2nd Edition), F. B. Hildebrand. Classic, fundamental treatment covers computation, approximation, interpolation, numerical differentiation and integration, other topics. 150 new problems. 669pp. 5⅜ x 8½.   65363-3

THE FUNCTIONS OF MATHEMATICAL PHYSICS, Harry Hochstadt. Comprehensive treatment of orthogonal polynomials, hypergeometric functions, Hill's equation, much more. Bibliography. Index. 322pp. 5⅜ x 8½. 65214-9

THREE PEARLS OF NUMBER THEORY, A. Y. Khinchin. Three compelling puzzles require proof of a basic law governing the world of numbers. Challenges concern van der Waerden's theorem, the Landau-Schnirelmann hypothesis and Mann's theorem, and a solution to Waring's problem. Solutions included. 64pp. 5⅜ x 8½. 40026-3

CALCULUS REFRESHER FOR TECHNICAL PEOPLE, A. Albert Klaf. Covers important aspects of integral and differential calculus via 756 questions. 566 problems, most answered. 431pp. 5⅜ x 8½. 20370-0

THE PHILOSOPHY OF MATHEMATICS: An Introductory Essay, Stephan Körner. Surveys the views of Plato, Aristotle, Leibniz & Kant concerning propositions and theories of applied and pure mathematics. Introduction. Two appendices. Index. 198pp. 5⅜ x 8½. 25048-2

INTRODUCTORY REAL ANALYSIS, A.N. Kolmogorov, S. V. Fomin. Translated by Richard A. Silverman. Self-contained, evenly paced introduction to real and functional analysis. Some 350 problems. 403pp. 5⅜ x 8½. 61226-0

APPLIED ANALYSIS, Cornelius Lanczos. Classic work on analysis and design of finite processes for approximating solution of analytical problems. Algebraic equations, matrices, harmonic analysis, quadrature methods, much more. 559pp. 5⅜ x 8½. 65656-X

AN INTRODUCTION TO ALGEBRAIC STRUCTURES, Joseph Landin. Superb self-contained text covers "abstract algebra": sets and numbers, theory of groups, theory of rings, much more. Numerous well-chosen examples, exercises. 247pp. 5⅜ x 8½. 65940-2

SPECIAL FUNCTIONS, N. N. Lebedev. Translated by Richard Silverman. Famous Russian work treating more important special functions, with applications to specific problems of physics and engineering. 38 figures. 308pp. 5⅜ x 8½. 60624-4

QUALITATIVE THEORY OF DIFFERENTIAL EQUATIONS, V. V. Nemytskii and V.V. Stepanov. Classic graduate-level text by two prominent Soviet mathematicians covers classical differential equations as well as topological dynamics and ergodic theory. Bibliographies. 523pp. 5⅜ x 8½. 65954-2

NUMBER THEORY AND ITS HISTORY, Oystein Ore. Unusually clear, accessible introduction covers counting, properties of numbers, prime numbers, much more. Bibliography. 380pp. 5⅜ x 8½. 65620-9

THEORY OF MATRICES, Sam Perlis. Outstanding text covering rank, nonsingularity and inverses in connection with the development of canonical matrices under the relation of equivalence, and without the intervention of determinants. Includes exercises. 237pp. 5⅜ x 8½. 66810-X

INTRODUCTION TO ANALYSIS, Maxwell Rosenlicht. Unusually clear, accessible coverage of set theory, real number system, metric spaces, continuous functions, Riemann integration, multiple integrals, more. Wide range of problems. Undergraduate level. Bibliography. 254pp. 5⅜ x 8½.   65038-3

MODERN NONLINEAR EQUATIONS, Thomas L. Saaty. Emphasizes practical solution of problems; covers seven types of equations. ". . . a welcome contribution to the existing literature...."–*Math Reviews.* 490pp. 5⅜ x 8½.   64232-1

MATRICES AND LINEAR ALGEBRA, Hans Schneider and George Phillip Barker. Basic textbook covers theory of matrices and its applications to systems of linear equations and related topics such as determinants, eigenvalues and differential equations. Numerous exercises. 432pp. 5⅜ x 8½.   66014-1

MATHEMATICS APPLIED TO CONTINUUM MECHANICS, Lee A. Segel. Analyzes models of fluid flow and solid deformation. For upper-level math, science and engineering students. 608pp. 5⅜ x 8½.   65369-2

ELEMENTS OF REAL ANALYSIS, David A. Sprecher. Classic text covers fundamental concepts, real number system, point sets, functions of a real variable, Fourier series, much more. Over 500 exercises. 352pp. 5⅜ x 8½.   65385-4

AN INTRODUCTION TO MATRICES, SETS AND GROUPS FOR SCIENCE STUDENTS, G. Stephenson. Concise, readable text introduces sets, groups, and most importantly, matrices to undergraduate students of physics, chemistry, and engineering. Problems. 164pp. 5⅜ x 8½.   65077-4

SET THEORY AND LOGIC, Robert R. Stoll. Lucid introduction to unified theory of mathematical concepts. Set theory and logic seen as tools for conceptual understanding of real number system. 496pp. 5⅜ x 8¼.   63829-4

TENSOR CALCULUS, J.L. Synge and A. Schild. Widely used introductory text covers spaces and tensors, basic operations in Riemannian space, non-Riemannian spaces, etc. 324pp. 5⅜ x 8¼.   63612-7

ORDINARY DIFFERENTIAL EQUATIONS, Morris Tenenbaum and Harry Pollard. Exhaustive survey of ordinary differential equations for undergraduates in mathematics, engineering, science. Thorough analysis of theorems. Diagrams. Bibliography. Index. 818pp. 5⅜ x 8½.   64940-7

INTEGRAL EQUATIONS, F. G. Tricomi. Authoritative, well-written treatment of extremely useful mathematical tool with wide applications. Volterra Equations, Fredholm Equations, much more. Advanced undergraduate to graduate level. Exercises. Bibliography. 238pp. 5⅜ x 8½.   64828-1

FOURIER SERIES, Georgi P. Tolstov. Translated by Richard A. Silverman. A valuable addition to the literature on the subject, moving clearly from subject to subject and theorem to theorem. 107 problems, answers. 336pp. 5⅜ x 8½.   63317-9

POPULAR LECTURES ON MATHEMATICAL LOGIC, Hao Wang. Noted logician's lucid treatment of historical developments, set theory, model theory, recursion theory and constructivism, proof theory, more. 3 appendixes. Bibliography. 1981 edition. ix + 283pp. 5⅜ x 8½.   67632-3

CALCULUS OF VARIATIONS, Robert Weinstock. Basic introduction covering isoperimetric problems, theory of elasticity, quantum mechanics, electrostatics, etc. Exercises throughout. 326pp. 5⅜ x 8½.   63069-2

THE CONTINUUM: A Critical Examination of the Foundation of Analysis, Hermann Weyl. Classic of 20th-century foundational research deals with the conceptual problem posed by the continuum. 156pp. 5⅜ x 8½.   67982-9

CHALLENGING MATHEMATICAL PROBLEMS WITH ELEMENTARY SOLUTIONS, A. M. Yaglom and I. M. Yaglom. Over 170 challenging problems on probability theory, combinatorial analysis, points and lines, topology, convex polygons, many other topics. Solutions. Total of 445pp. 5⅜ x 8½. Two-vol. set.
Vol. I: 65536-9   Vol. II: 65537-7

A SURVEY OF NUMERICAL MATHEMATICS, David M. Young and Robert Todd Gregory. Broad self-contained coverage of computer-oriented numerical algorithms for solving various types of mathematical problems in linear algebra, ordinary and partial, differential equations, much more. Exercises. Total of 1,248pp. 5⅜ x 8½. Two volumes.   Vol. I: 65691-8   Vol. II: 65692-6

INTRODUCTION TO PARTIAL DIFFERENTIAL EQUATIONS WITH APPLICATIONS, E. C. Zachmanoglou and Dale W. Thoe. Essentials of partial differential equations applied to common problems in engineering and the physical sciences. Problems and answers. 416pp. 5⅜ x 8½.   65251-3

THE THEORY OF GROUPS, Hans J. Zassenhaus. Well-written graduate-level text acquaints reader with group-theoretic methods and demonstrates their usefulness in mathematics. Axioms, the calculus of complexes, homomorphic mapping, p-group theory, more. Many proofs shorter and more transparent than older ones. 276pp. 5⅜ x 8½.   40922-8

DISTRIBUTION THEORY AND TRANSFORM ANALYSIS: An Introduction to Generalized Functions, with Applications, A. H. Zemanian. Provides basics of distribution theory, describes generalized Fourier and Laplace transformations. Numerous problems. 384pp. 5⅜ x 8½.   65479-6

Math–Decision Theory, Statistics, Probability

ELEMENTARY DECISION THEORY, Herman Chernoff and Lincoln E. Moses. Clear introduction to statistics and statistical theory covers data processing, probability and random variables, testing hypotheses, much more. Exercises. 364pp. 5⅜ x 8½.   65218-1

STATISTICS MANUAL, Edwin L. Crow et al. Comprehensive, practical collection of classical and modern methods prepared by U.S. Naval Ordnance Test Station. Stress on use. Basics of statistics assumed. 288pp. 5⅜ x 8½.   60599-X

SOME THEORY OF SAMPLING, William Edwards Deming. Analysis of the problems, theory and design of sampling techniques for social scientists, industrial managers and others who find statistics important at work. 61 tables. 90 figures. xvii +602pp. 5⅜ x 8½.   64684-X

STATISTICAL ADJUSTMENT OF DATA, W. Edwards Deming. Introduction to basic concepts of statistics, curve fitting, least squares solution, conditions without parameter, conditions containing parameters. 26 exercises worked out. 271pp. 5⅜ x 8½.   64685-8

LINEAR PROGRAMMING AND ECONOMIC ANALYSIS, Robert Dorfman, Paul A. Samuelson and Robert M. Solow. First comprehensive treatment of linear programming in standard economic analysis. Game theory, modern welfare economics, Leontief input-output, more. 525pp. 5⅜ x 8½.   65491-5

DICTIONARY/OUTLINE OF BASIC STATISTICS, John E. Freund and Frank J. Williams. A clear concise dictionary of over 1,000 statistical terms and an outline of statistical formulas covering probability, nonparametric tests, much more. 208pp. 5⅜ x 8½.   66796-0

PROBABILITY: An Introduction, Samuel Goldberg. Excellent basic text covers set theory, probability theory for finite sample spaces, binomial theorem, much more. 360 problems. Bibliographies. 322pp. 5⅜ x 8½.   65252-1

GAMES AND DECISIONS: Introduction and Critical Survey, R. Duncan Luce and Howard Raiffa. Superb nontechnical introduction to game theory, primarily applied to social sciences. Utility theory, zero-sum games, n-person games, decision-making, much more. Bibliography. 509pp. 5⅜ x 8½.   65943-7

FIFTY CHALLENGING PROBLEMS IN PROBABILITY WITH SOLUTIONS, Frederick Mosteller. Remarkable puzzlers, graded in difficulty, illustrate elementary and advanced aspects of probability. Detailed solutions. 88pp. 5⅜ x 8½.   65355-2

PROBABILITY THEORY: A Concise Course, Y. A. Rozanov. Highly readable, self-contained introduction covers combination of events, dependent events, Bernoulli trials, etc. 148pp. 5⅜ x 8¼.   63544-9

STATISTICAL METHOD FROM THE VIEWPOINT OF QUALITY CONTROL, Walter A. Shewhart. Important text explains regulation of variables, uses of statistical control to achieve quality control in industry, agriculture, other areas. 192pp. 5⅜ x 8½.   65232-7

THE COMPLEAT STRATEGYST: Being a Primer on the Theory of Games of Strategy, J. D. Williams. Highly entertaining classic describes, with many illustrated examples, how to select best strategies in conflict situations. Prefaces. Appendices. 268pp. 5⅜ x 8½.   25101-2

Math–Geometry and Topology

ELEMENTARY CONCEPTS OF TOPOLOGY, Paul Alexandroff. Elegant, intuitive approach to topology from set-theoretic topology to Betti groups; how concepts of topology are useful in math and physics. 25 figures. 57pp. 5⅜ x 8½. 60747-X

COMBINATORIAL TOPOLOGY, P. S. Alexandrov. Clearly written, well-organized, three-part text begins by dealing with certain classic problems without using the formal techniques of homology theory and advances to the central concept, the Betti groups. Numerous detailed examples. 654pp. 5⅜ x 8½. 40179-0

EXPERIMENTS IN TOPOLOGY, Stephen Barr. Classic, lively explanation of one of the byways of mathematics. Klein bottles, Moebius strips, projective planes, map coloring, problem of the Koenigsberg bridges, much more, described with clarity and wit. 43 figures. 210pp. 5⅜ x 8½. 25933-1

CONFORMAL MAPPING ON RIEMANN SURFACES, Harvey Cohn. Lucid, insightful book presents ideal coverage of subject. 334 exercises make book perfect for self-study. 55 figures. 352pp. 5⅜ x 8¼. 64025-6

THE GEOMETRY OF RENÉ DESCARTES, René Descartes. The great work founded analytical geometry. Original French text, Descartes's own diagrams, together with definitive Smith-Latham translation. 244pp. 5⅜ x 8½. 60068-8

THE THIRTEEN BOOKS OF EUCLID'S ELEMENTS, translated with introduction and commentary by Sir Thomas L. Heath. Definitive edition. Textual and linguistic notes, mathematical analysis. 2,500 years of critical commentary. Unabridged. 1,414pp. 5⅜ x 8½. Three-vol. set.
Vol. I: 60088-2 Vol. II: 60089-0 Vol. III: 60090-4

GEOMETRY OF COMPLEX NUMBERS, Hans Schwerdtfeger. Illuminating, widely praised book on analytic geometry of circles, the Moebius transformation, and two-dimensional non-Euclidean geometries. 200pp. 5⅜ x 8¼. 63830-8

DIFFERENTIAL GEOMETRY, Heinrich W. Guggenheimer. Local differential geometry as an application of advanced calculus and linear algebra. Curvature, transformation groups, surfaces, more. Exercises. 62 figures. 378pp. 5⅜ x 8½. 63433-7

CURVATURE AND HOMOLOGY: Enlarged Edition, Samuel I. Goldberg. Revised edition examines topology of differentiable manifolds; curvature, homology of Riemannian manifolds; compact Lie groups; complex manifolds; curvature, homology of Kaehler manifolds. New Preface. Four new appendixes. 416pp. 5⅜ x 8½. 40207-X

TOPOLOGY, John G. Hocking and Gail S. Young. Superb one-year course in classical topology. Topological spaces and functions, point-set topology, much more. Examples and problems. Bibliography. Index. 384pp. 5⅜ x 8¼. 65676-4

LECTURES ON CLASSICAL DIFFERENTIAL GEOMETRY, Second Edition, Dirk J. Struik. Excellent brief introduction covers curves, theory of surfaces, fundamental equations, geometry on a surface, conformal mapping, other topics. Problems. 240pp. 5⅜ x 8½. 65609-8

Math–History of

A SHORT ACCOUNT OF THE HISTORY OF MATHEMATICS, W. W. Rouse Ball. One of clearest, most authoritative surveys from the Egyptians and Phoenicians through 19th-century figures such as Grassman, Galois, Riemann. Fourth edition. 522pp. 5⅜ x 8½. 20630-0

THE HISTORY OF THE CALCULUS AND ITS CONCEPTUAL DEVELOPMENT, Carl B. Boyer. Origins in antiquity, medieval contributions, work of Newton, Leibniz, rigorous formulation. Treatment is verbal. 346pp. 5⅜ x 8½. 60509-4

THE HISTORICAL ROOTS OF ELEMENTARY MATHEMATICS, Lucas N. H. Bunt, Phillip S. Jones, and Jack D. Bedient. Fundamental underpinnings of modern arithmetic, algebra, geometry and number systems derived from ancient civilizations. 320pp. 5⅜ x 8½. 25563-8

A HISTORY OF MATHEMATICAL NOTATIONS, Florian Cajori. This classic study notes the first appearance of a mathematical symbol and its origin, the competition it encountered, its spread among writers in different countries, its rise to popularity, its eventual decline or ultimate survival. Original 1929 two-volume edition presented here in one volume. xxviii+820pp. 5⅜ x 8½. 67766-4

GAMES, GODS & GAMBLING: A History of Probability and Statistical Ideas, F. N. David. Episodes from the lives of Galileo, Fermat, Pascal, and others illustrate this fascinating account of the roots of mathematics. Features thought-provoking references to classics, archaeology, biography, poetry. 1962 edition. 304pp. 5⅜ x 8½. (Available in U.S. only.) 40023-9

OF MEN AND NUMBERS: The Story of the Great Mathematicians, Jane Muir. Fascinating accounts of the lives and accomplishments of history's greatest mathematical minds–Pythagoras, Descartes, Euler, Pascal, Cantor, many more. Anecdotal, illuminating. 30 diagrams. Bibliography. 256pp. 5⅜ x 8½. 28973-7

HISTORY OF MATHEMATICS, David E. Smith. Nontechnical survey from ancient Greece and Orient to late 19th century; evolution of arithmetic, geometry, trigonometry, calculating devices, algebra, the calculus. 362 illustrations. 1,355pp. 5⅜ x 8½. Two-vol. set. Vol. I: 20429-4 Vol. II: 20430-8

A CONCISE HISTORY OF MATHEMATICS, Dirk J. Struik. The best brief history of mathematics. Stresses origins and covers every major figure from ancient Near East to 19th century. 41 illustrations. 195pp. 5⅜ x 8½. 60255-9

Physics

OPTICAL RESONANCE AND TWO-LEVEL ATOMS, L. Allen and J. H. Eberly. Clear, comprehensive introduction to basic principles behind all quantum optical resonance phenomena. 53 illustrations. Preface. Index. 256pp. 5⅜ x 8½.   65533-4

ULTRASONIC ABSORPTION: An Introduction to the Theory of Sound Absorption and Dispersion in Gases, Liquids and Solids, A. B. Bhatia. Standard reference in the field provides a clear, systematically organized introductory review of fundamental concepts for advanced graduate students, research workers. Numerous diagrams. Bibliography. 440pp. 5⅜ x 8½.   64917-2

QUANTUM THEORY, David Bohm. This advanced undergraduate-level text presents the quantum theory in terms of qualitative and imaginative concepts, followed by specific applications worked out in mathematical detail. Preface. Index. 655pp. 5⅜ x 8½.   65969-0

ATOMIC PHYSICS (8th edition), Max Born. Nobel laureate's lucid treatment of kinetic theory of gases, elementary particles, nuclear atom, wave-corpuscles, atomic structure and spectral lines, much more. Over 40 appendices, bibliography. 495pp. 5⅜ x 8½.   65984-4

AN INTRODUCTION TO HAMILTONIAN OPTICS, H. A. Buchdahl. Detailed account of the Hamiltonian treatment of aberration theory in geometrical optics. Many classes of optical systems defined in terms of the symmetries they possess. Problems with detailed solutions. 1970 edition. xv + 360pp. 5⅜ x 8½.   67597-1

THIRTY YEARS THAT SHOOK PHYSICS: The Story of Quantum Theory, George Gamow. Lucid, accessible introduction to influential theory of energy and matter. Careful explanations of Dirac's anti-particles, Bohr's model of the atom, much more. 12 plates. Numerous drawings. 240pp. 5⅜ x 8½.   24895-X

ELECTRONIC STRUCTURE AND THE PROPERTIES OF SOLIDS: The Physics of the Chemical Bond, Walter A. Harrison. Innovative text offers basic understanding of the electronic structure of covalent and ionic solids, simple metals, transition metals and their compounds. Problems. 1980 edition. 582pp. 6⅛ x 9¼.   66021-4

HYDRODYNAMIC AND HYDROMAGNETIC STABILITY, S. Chandrasekhar. Lucid examination of the Rayleigh-Benard problem; clear coverage of the theory of instabilities causing convection. 704pp. 5⅜ x 8¼.   64071-X

INVESTIGATIONS ON THE THEORY OF THE BROWNIAN MOVEMENT, Albert Einstein. Five papers (1905–8) investigating dynamics of Brownian motion and evolving elementary theory. Notes by R. Fürth. 122pp. 5⅜ x 8½.   60304-0

THE PHYSICS OF WAVES, William C. Elmore and Mark A. Heald. Unique overview of classical wave theory. Acoustics, optics, electromagnetic radiation, more. Ideal as classroom text or for self-study. Problems. 477pp. 5⅜ x 8½.   64926-1

PHYSICAL PRINCIPLES OF THE QUANTUM THEORY, Werner Heisenberg. Nobel Laureate discusses quantum theory, uncertainty, wave mechanics, work of Dirac, Schroedinger, Compton, Wilson, Einstein, etc. 184pp. 5⅜ x 8½. 60113-7

ATOMIC SPECTRA AND ATOMIC STRUCTURE, Gerhard Herzberg. One of best introductions; especially for specialist in other fields. Treatment is physical rather than mathematical. 80 illustrations. 257pp. 5⅜ x 8½. 60115-3

AN INTRODUCTION TO STATISTICAL THERMODYNAMICS, Terrell L. Hill. Excellent basic text offers wide-ranging coverage of quantum statistical mechanics, systems of interacting molecules, quantum statistics, more. 523pp. 5⅜ x 8½. 65242-4

THEORETICAL PHYSICS, Georg Joos, with Ira M. Freeman. Classic overview covers essential math, mechanics, electromagnetic theory, thermodynamics, quantum mechanics, nuclear physics, other topics. First paperback edition. xxiii + 885pp. 5⅜ x 8½. 65227-0

PROBLEMS AND SOLUTIONS IN QUANTUM CHEMISTRY AND PHYSICS, Charles S. Johnson, Jr. and Lee G. Pedersen. Unusually varied problems, detailed solutions in coverage of quantum mechanics, wave mechanics, angular momentum, molecular spectroscopy, more. 280 problems plus 139 supplementary exercises. 430pp. 6½ x 9¼. 65236-X

THEORETICAL SOLID STATE PHYSICS, Vol. 1: Perfect Lattices in Equilibrium; Vol. II: Non-Equilibrium and Disorder, William Jones and Norman H. March. Monumental reference work covers fundamental theory of equilibrium properties of perfect crystalline solids, non-equilibrium properties, defects and disordered systems. Appendices. Problems. Preface. Diagrams. Index. Bibliography. Total of 1,301pp. 5⅜ x 8½. Two volumes. Vol. I: 65015-4 Vol. II: 65016-2

A TREATISE ON ELECTRICITY AND MAGNETISM, James Clerk Maxwell. Important foundation work of modern physics. Brings to final form Maxwell's theory of electromagnetism and rigorously derives his general equations of field theory. 1,084pp. 5⅜ x 8½. Two-vol. set. Vol. I: 60636-8 Vol. II: 60637-6

OPTICKS, Sir Isaac Newton. Newton's own experiments with spectroscopy, colors, lenses, reflection, refraction, etc., in language the layman can follow. Foreword by Albert Einstein. 532pp. 5⅜ x 8½. 60205-2

THEORY OF ELECTROMAGNETIC WAVE PROPAGATION, Charles Herach Papas. Graduate-level study discusses the Maxwell field equations, radiation from wire antennas, the Doppler effect and more. xiii + 244pp. 5⅜ x 8½. 65678-5

INTRODUCTION TO QUANTUM MECHANICS With Applications to Chemistry, Linus Pauling & E. Bright Wilson, Jr. Classic undergraduate text by Nobel Prize winner applies quantum mechanics to chemical and physical problems. Numerous tables and figures enhance the text. Chapter bibliographies. Appendices. Index. 468pp. 5⅜ x 8½. 64871-0

CATALOG OF DOVER BOOKS

METHODS OF THERMODYNAMICS, Howard Reiss. Outstanding text focuses on physical technique of thermodynamics, typical problem areas of understanding, and significance and use of thermodynamic potential. 1965 edition. 238pp. 5⅜ x 8½. 69445-3

TENSOR ANALYSIS FOR PHYSICISTS, J. A. Schouten. Concise exposition of the mathematical basis of tensor analysis, integrated with well-chosen physical examples of the theory. Exercises. Index. Bibliography. 289pp. 5⅜ x 8½. 65582-2

RELATIVITY IN ILLUSTRATIONS, Jacob T. Schwartz. Clear nontechnical treatment makes relativity more accessible than ever before. Over 60 drawings illustrate concepts more clearly than text alone. Only high school geometry needed. Bibliography. 128pp. 6⅛ x 9¼. 25965-X

THE ELECTROMAGNETIC FIELD, Albert Shadowitz. Comprehensive undergraduate text covers basics of electric and magnetic fields, builds up to electromagnetic theory. Also related topics, including relativity. Over 900 problems. 768pp. 5⅝ x 8¼. 65660-8

GREAT EXPERIMENTS IN PHYSICS: Firsthand Accounts from Galileo to Einstein, edited by Morris H. Shamos. 25 crucial discoveries: Newton's laws of motion, Chadwick's study of the neutron, Hertz on electromagnetic waves, more. Original accounts clearly annotated. 370pp. 5⅜ x 8½. 25346-5

RELATIVITY, THERMODYNAMICS AND COSMOLOGY, Richard C. Tolman. Landmark study extends thermodynamics to special, general relativity; also applications of relativistic mechanics, thermodynamics to cosmological models. 501pp. 5⅜ x 8½. 65383-8

LIGHT SCATTERING BY SMALL PARTICLES, H. C. van de Hulst. Comprehensive treatment including full range of useful approximation methods for researchers in chemistry, meteorology and astronomy. 44 illustrations. 470pp. 5⅜ x 8½. 64228-3

STATISTICAL PHYSICS, Gregory H. Wannier. Classic text combines thermodynamics, statistical mechanics and kinetic theory in one unified presentation of thermal physics. Problems with solutions. Bibliography. 532pp. 5⅜ x 8½. 65401-X

Paperbound unless otherwise indicated. Available at your book dealer, online at **www.doverpublications.com**, or by writing to Dept. GI, Dover Publications, Inc., 31 East 2nd Street, Mineola, NY 11501. For current price information or for free catalogues (please indicate field of interest), write to Dover Publications or log on to **www.doverpublications.com** and see every Dover book in print. Dover publishes more than 500 books each year on science, elementary and advanced mathematics, biology, music, art, literary history, social sciences, and other areas.